PROBLEMS IN
HEAT

PROBLEMS
IN
HEAT

By

M.S. Yadav

ANMOL PUBLICATIONS PVT. LTD.
NEW DELHI - 110 002 (INDIA)

ANMOL PUBLICATIONS PVT. LTD.
4374/4B, Ansari Road, Daryaganj
New Delhi - 110 002

Problems in Heat

First Edition, 2000
ISBN 81-261-0549-6

PRINTED IN INDIA

Published by J.L. Kumar for Anmol Publications Pvt. Ltd., New Delhi - 110 002 and Printed at Mehra Offset Press, Delhi.

Contents

Preface

Problems in Heat includes problems and their solutions on heat, categorised in nineteen chapters. This division of book into chapters has been done to present the subject matter in a comprehensive but easily understandable manner. It has also helped to avoid repetition of the subject matter. Most of the topics included in this form a part of syllabus for B.Sc. and M.Sc. students of the Indian Universities. The topics included are Thermometry; Expansion of Solids; Expansion of Liquids; Expansion of Gases; First Law of Thermodynamics; Second Law of Thermodynamics; Production of Very Low Temperatures; The Mechanical Equivalent of Heat; Entropy; Thermodynamic Relationships; Classical Statistics; Kinetic Theory; Behaviour of Real Gases; Change of State; Specific Heats of Gases; Thermal Radiation; Conduction and Convection, etc.

An attempt has been made to present the subject matter in such a form that it is liked by students and teachers alike.

I am thankful to a number of authors whose works have been freely consulted while preparing this book. I am thankful to M/s. Anmol Publications Pvt. Ltd., New Delhi for bringing out the book promptly.

I shall thankfully accept the suggestions for modification and improvement in the book.

—M.S. Yadav

1

Thermometry

Problem 1 : Calculate the value of R (Universal gas constant) in ergs per degree centigrade per gram-mol and calories per degree centigrade per gram-mol.

Solution : These can be calculated by use of gas equation.

for 1 mol of a gas, we have

$$PV = RT$$

For any gas at N.T.P. the gram molecular volume is 22.4 litres.

Substituting numerical value of P, V and T is the above equation, we have

$$(76 \times 13.595 \times 980) \times (22400) = R \times 273$$

$$\therefore \; R = \frac{76 \times 13.595 \times 980 \times 22400}{273}$$

$= 8.313 \times 10^7$ ergs per degree centegrade per gram-mol

Since 1 caloric $= 4.2 \times 10^7$ ergs

$$\therefore R = \frac{8.313 \times 10^7}{4.2 \times 10^7}$$

= 1.986 calories per degree centigrade per gram-mol

Problem 2 : The pressure of hydrogen in a constant volume gas thermo meter is 74.0 cm mercury at the temperature of melting ice; 108.4 cm at the temperature of steam at normal pressure, and 87.2 cm at the temperature 't' of a liquid. Find the value of 't'.

Solution : The temperature 't' on the constant-volume gas thermometer is calculated from the relation.

$$\frac{t}{100} = \frac{p_t - p_0}{p_{100} - p_0}$$

so we get 't' by substituting the given values,

$$\frac{t}{100} = \frac{87.2 - 74.0}{108.4 - 74.0}$$

$$\text{or } t = 38.4° \text{ C}$$

Problem 3 : If the values of R_0 and R_{100} in a platinum resistance thermometer are 6.284 ohms and 7.321 ohms respectively and the R_t in 6.726 ohm when the thermometer is placed in warm water. What is per temperature (t_p) of warm water?

Solution : The value of t_p can be calculated by using the relation

$$\frac{t_p}{100} = \frac{R_t - R_0}{R_{100} - R_0}$$

Substituting various values in the relation, we get

$$\frac{t_p}{100} = \frac{6.726 - 6.284}{7.321 - 6.284}$$

$$\text{or } t_p = 42.6° \text{ C}$$

Problem 4 : The following readings were obtained with a constant-volume air thermometer

	Level of mercury in closed limb (mm)	Level of mercury in open limb (mm)
Bulb in melting ice	136	112
Bulb in steam at 76 mm pressure	136	390
Bulb at room temp	136	160

Calculate the room temperature.

Solution : The temperature 't' of the room is given by

$$\frac{t}{100} = \frac{p_t - p_0}{p_{100} - p_0}$$

Now we have; $p_0 = H - (13.6 - 11.2)$ cm

where H is the barometric pressure in cm. of mercury.

$$p_t = H + (16.0 - 13.6) \text{ cm}$$

$$p_{100} = H + (39.0 - 13.6) \text{ cm}$$

Hence $P_t - p_0 = 4.8$ cm

and $p_{100} - p_0 = 27.8$ cm

$$\frac{t}{100} = \frac{4.8}{27.8}$$

$$\therefore \frac{t}{100} = \frac{4.8}{27.8} \times 100 = 17.3°\text{C}$$

Problem 5 **:** In a mercury thermometer the volumes of the mercury at 0°C and 100° C are 0.400 cc and 0.432 cc respectively. Calculate the temperature when the volume of the mercury is (i) 0.408 cc (ii) 0.428 cc (iii) 0.390 cc.

Solution : The temperature is given by the relation

$$\frac{t}{100} = \frac{v_t - v_0}{v_{100} - v_0}$$

Given : $v_0 = 0.400$ cc, $v_{100} = 0.432$ cc;

(i) $\therefore v_t - v_0 = 0.408 - 0.400 = 0.008$ cc.

$v_{100} - v_0 = 0.432 - 0.400 = 0.032$ cc.

$$\frac{t}{100} = \frac{0.008}{0.032}$$

$$\therefore \quad t = \frac{0.008}{0.032}$$

$$\text{or } t = \frac{0.008}{0.032} \times 100 = 25°\text{C}$$

(ii) Proceed in same way. Ans – 87.5° C

(iii) Proceed in same way. Ans – 31.3° C

Problem 6 : Resistance of a platinum resistance thermometer at triple point of water is 26.22 Ω. When kept in boiling glycerol, this resistance became 54.41 Ω. Find out the boiling point of glycesol as this scale. Compare it with standard value of boiling point of glycerol (563.2 K) and explain the difference if any.

Solution : Using the relation,

$$\frac{t_p}{273.16} = \frac{R_t - R_0}{R_{100} - R_0}, \text{ we get}$$

$$t_p = 273.16 \times \frac{54.41}{26.22}$$

$$= 567.0 \text{ degrees.}$$

It is ligher than the standard value (563.2K). The difference is due to the fact that change in resistance of a platinum wire with temperature is not exactly similar to the change in pressure of an ideal gas.

Problem 7 : A cylindrical vessel contains an ideal gas at a temperature of 27°C and a pressure of atmosphere. The volume of the gas is ten liters. The gas is compressed with the help of a piston to half of its original volume. The temperature rises to 47°C. What is the new pressure? How many moles of gas are there?

Solution : In this process there is no change in the mass of the gas so we can use the relation

$$PV = nRT$$

$$\text{or } \frac{P_1V_1}{T_1} = \frac{P_2V_2}{T_2}$$

From the given data we have

$P_1 = 1$ atmo	$P_2 = ?$
$V_1 = 10$ litres	$V_2 = 5$ litres
$T_1 = 27 + 273 = 300$ K	$T_2 = 47 + 273 = 320$ K

$$\therefore \frac{P_2 \times 5}{320} = \frac{1 \times 10}{300}$$

$$\text{or} \quad P_2 = \frac{320 \times 10}{300 \times 5} = 2.13 \text{ atmospheres.}$$

Again from PV = nRT, we get

$$n = \frac{PV}{RT}$$

$$\text{or} \quad n = \frac{P_1 V_1}{RT_1}$$

$$\text{or} \quad n = \frac{1.013 \times 10^5\ Nm^{-2} \times 10 \times 10^{-3}\ m^3}{8.314\ JK^{-1}\ mol^{-1} \times 300\ K}$$

$$= \frac{1.013 \times 10}{8.314 \times 3} \text{ moles}$$

$$= 0.406 \text{ mole}$$

2

Expansion of Solids

Problem 1 : An iron wire of diameter 0.4 mm is heated uniformly to 300°C and rigidly clamped at its ends while hot. Find the pull exerted on the clamps when it has cooled to 20°C. Coefficient of linear expansion of iron is 1×10^{-5} and Young's modulus of iron is 2.1×10^{12} dynes sq. cm.

Solution : Let the original length at 20°C = l

$$\text{Increase in length} = l . \alpha . t$$

$$= l \times 1 \times 10^{-5} . (300 - 20)$$

$$\therefore \text{Strain} = \frac{l \times 10^{-5} \times 280}{l}$$

$$= 280 \times 10^{-5}$$

Diameter of the wire = 0.4 mm = 0.04 cm.

$$\therefore \quad \text{Area of cross section} = \frac{22}{7} \times 0.02 \times 0.02$$

If F is the full exerted, then

$$\text{Stress} = \frac{F}{A}$$

$$= \frac{F \times 7}{22 \times 0.02 \times 0.02}$$

$$\text{Since Young's modulus} = \frac{\text{Stress}}{\text{Strain}}$$

$$\therefore\ 2.1 \times 10^{12} = \frac{F \times 7}{22 \times 0.02 \times 0.02 \times 280 \times 10^{-5}}$$

$$\text{or } F = 7.392 \times 10^{6} \text{ dynes.}$$

Problem 2 : In an experiment using a light of wave length, $\alpha = 0.00005896$ cm and a plate of rock-salt crystal 5 mm Hick, it was found heat a displacement of 9 interference band took place when the temperature was increased from 10° to 50°C. Find the coefficient of linear expansion of rock-salt

Solution : The increase in length $= x \cdot \frac{\alpha}{2}$

$$= 9 \times \frac{0.00005896}{2}$$

$$= 0.000261 \text{ cm.}$$

$\therefore$ Coefficient of linear expansion of rock-salt $= \frac{0.000261}{0.5 \times 40}$

$$= 0.00001305$$

$$= 13.05 \times 10^{-6}$$

Problem 3 : Calculate how much a pendulum clock will lose in a month of 30 days if the average temperature is 18°C. Assuming that the clock is accurate at 8°C, and has then a period of exactly 2 sec.

Solution : If the length of the pendulum is l_1, at 8°C and it increases to l_2 at 18°C.

$$\text{Then } \frac{T_2}{T_1} = \sqrt{\frac{l_2}{l_1}}$$

where T_1 and T_2 are respective time periods at 8°C and 18°C

$$\text{But } l_2 = l_1\ [1 + (18 - 8)\ \alpha\ ]$$

where α is the mean coefficient of expansion of the pendulum wire.

$$\therefore\ \frac{T_2}{T_1} = \sqrt{\frac{l_1\,(1 + 10\alpha)}{l_1}} = \sqrt{1 + 10\alpha}$$

But $T_1 = 2$ sec; $\alpha = 0.000011$ per °C, if the wire is made of iron

$$\therefore \frac{T_2}{2} = \sqrt{1 + 10 \times .000011} = \sqrt{1.00011}$$

or $T_2 = 2 \times \sqrt{1.00011} = 2.00011$ sec.

$\therefore$ in 2.00011 sec clock register 2 sec.

or in 2.00011 sec clock loses 0.00011 sec

or in 1 sec clock loses $\simeq \frac{0.00011}{2}$ sec.

$$\text{So time lost in 30 days} = \frac{0.00011}{2} \times 30 \times 24 \times 3600$$

$$= 143 \text{ sec}$$

$$= 2 \text{ min } 23 \text{ sec.}$$

Problem 4 : The barometric height, as read at 12°C, by a brass scale correct at 0°C, is 76.52 cm. What is the actual height of the mercury column at 12°C? (α for brass = 0.000019 per degree centigrade)

Solution : The length 76.52 cm on brass scale is correct only at 0°C. The brass scale expands when the temperature rises to 12°C and the new length l_2 of 76.52 cm of brass at 0°C is given by

$$l_2 = l_1 [1 + \alpha \ (t_2 - t_1)]$$

$$= 76.52 [1 + 0.000019 (12 - 0)]$$

$$= 76.52 \times 1.000228$$

$$= 76.54$$

Thus the actual height of mercury column at 12°C is 76.54 cm.

Problem 5 : The diagram shows an iron wire AB stretched inside a rigid brass frame work and rigidty attached to it at both ends, A and

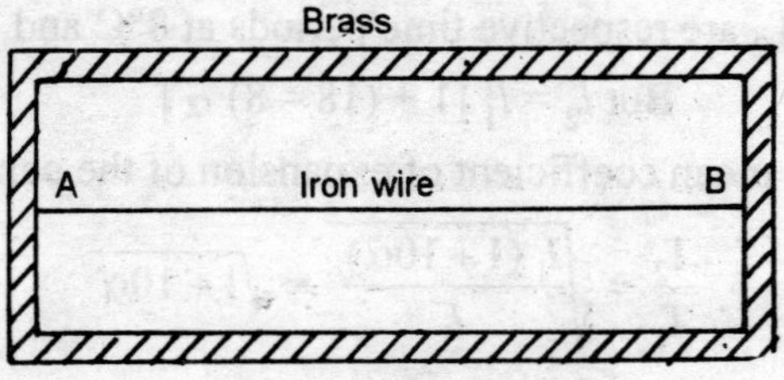

Fig.

B. The length of AB at 0°C is 300 cm and the diameter of the wire is 0.6 mm. What extra tension will be set up in the stretched wire when the temperature of the system is raised to 40°C? (α for iron = 0.000012 per °C; α for brass = 0.000018 per °C; Youngs modulus for iron = 2.1×10^{12} dynes per sq. cm).

Solution :

$$\text{Youngs Modulus } E = \frac{\text{stress}}{\text{strain}} = \frac{F/A}{e/l}$$

$$\therefore\ F = \frac{EAC}{l}$$

on heating the brass expands to a length 300 (1 + 40 α)

$$= 300\,(1 + 40 \times 0.000018) = 300.216 \text{ cm.}$$

The iron wire would have normally expanded to

$$= 300\,(1 + 40 \times 0.000012) = 300.144 \text{ cm.}$$

Hence the extenstion (e) due to its attachment of the brass

$$= (300.216 - 300.144) \text{ cm} = 0.072 \text{ cm.}$$

The original length of the iron = 300.144

$$\therefore \text{ Extra tension in wire F} = \frac{EAe}{l}$$

$$= 2.1 \times 10^{12} \times (\pi \times 0.03^2) \times \frac{0.072}{300.144}$$

$$= 14.24 \times 10^5 \text{ dynes.}$$

3

Expansion of Liquids

Problem 1 : The density of mercury at 20°C is 13.55 g cm^{-3}. At what temperature will his figure be in error by 1%? (The mean coefficient of expansion of mercury is 182×10^{-6} per °C)

Solution : Suppose per error is 1% at a temp. t°C.

Then at his temperature the density is less than 13.55 by $\frac{1}{100}$ of 13.55

$\therefore$ density at t°C(d_1) = 13.55 – 0.1345 = 13.4145 g cm^{-3}

density at 20°C(d_2) = 13.55 gcm^{-3}

$$\therefore \frac{d_1}{d_2} = 1 + \gamma\,(t_2 - t_1)$$

$$\therefore \frac{13.55}{13.4145} = 1 + 0.000182\,(t - 20)$$

$$\text{or } t - 20 = \frac{13.55 - 13.4145}{13.4145 \times 0.000182}$$

$$= 55.6$$

$$\therefore t = 75.6°C$$

Problem 2 : When a glass weight thermometer filled with mercury at 0°C is heated to 100°C, 7.785 g of mercury overflow and 450.0g remain in the thermometer. Using glycerine in place of mercury, the corresponding figures are 2.173 g and 41.00 g. The coefficient of expansion of mercury is 0.000183 per °C. Determine (a) The coefficient of linear expansion of the glass (b) The coeficient of volume expansion of glycerine.

Solution :

(a) Apparent coefficient of mercury

$$\gamma_{app} = \frac{\text{mass expelled}}{\text{mass left} \times \text{rise in temp.}}$$

$$= \frac{7.785}{450 \times 100} = 0.000173 \text{ per } ^\circ\text{C}$$

But the absolute coefficient of mercury

$\gamma = \gamma_{app} + C$, where C is cubical coefficient of glass.

$$\therefore C = \gamma - \gamma_{app}$$

$$= 0.000183 - 0.000173$$

$$= 0.000010/^\circ\text{C}$$

$$\therefore \text{Coefficient of linear expansion of glass} = \frac{1}{3} \times C$$

$$= \frac{1}{3} \times 0.000010 / ^\circ\text{C}$$

$$= 0.0000033/^\circ\text{C}.$$

(b) Apparent coefficient γ_{app} of glycerine

$$\frac{\text{mass expelled}}{\text{mass left} \times \text{rise in temp}}$$

$$= \frac{2.173}{41.00 \times 100}$$

$$= 0.000530/^\circ\text{C}$$

$$\therefore \text{absolute coefficient of glycerine} = \gamma_{app} + c$$

$$= 0.000530 + 0.000010$$

$$= 0.000540/^\circ\text{C}.$$

Problem 3 : If l is the coefficient of linear expansion of a solid, explain carefully why $3l$ may be used as coefficient of cubical expansion.

Solution : When a solid or glass vessel, becomes hot, the volume changes as well as the length. Suppose that a *hollow cube* has each side of length l_0 at 0°C and is heated to a temperature of t°C. If α is the coefficient of linear expansion, each side expands to $l_0(1 + \alpha t)$. The new volume of the cube is thus

$$l_0^3\,(1+\alpha t)^3 = l_0^3\,(1+3\alpha t+3\alpha^2 t^2+\alpha^3 t^3)$$

and subtracting the original volume l_0^3

$$\text{increase in volume} = l_0^3\,(3\alpha t+3\alpha^2 t^2+\alpha^3 t^3)$$

The coefficient of cubical expansion of solid (γ) is "the increase in volume of unit volume at 0°C, when the temperature increases by 1°C."

$$\gamma = l_0^3\,(3\alpha t+3\alpha^2 t^2+\alpha^3 t^3)\,/\,(l_0^3\times t)$$

$$= 3\alpha+3\alpha^2 t+\alpha^3 t^2$$

But $(3\alpha^2 t+\alpha^3 t^2)$ is very small compared to 3α, since α is very small

$$\therefore\ \gamma = 3\alpha$$

Problem 4: The height of a mercury barometer read with a steel scale is 754 mm at 20°C. What will it read at 0°C? (Coefficient of linear expansion of steel is 0.000012 per °C; coefficient of cubical expansion of mercury is 0.000182 per °C)

Solution : Suppose that the steel scale reads correctly at 0°C. Then at 20°C, the reading of 754 mm on the steel scale is smaller than the true length, h_t, which is given by

$$h_t = 754\,(1\ +\ 0.000012\times 20)$$

Let h_0 = the required height of the barometer at 0°C Then, since the pressure at 0°C and 20°C, are the same,

$$h_0 d_0 g\ =\ h_t d_t g$$

where d_0, d_t are the densities of mercury at 0°C, 20°C respectively.

$$\therefore\ h_0 = h_t\,.\frac{d_t}{d_0} = \frac{h_t}{1+20\gamma}$$

since $\frac{d_t}{d_0} = \frac{1}{1 + \gamma t}$ where γ is the cubical coefficient of mercury 0.000182 per °C. But

$$h_t = 754\,(1 + 0.000012 \times 20)$$

$$\therefore\ h_0 = \frac{754\,(1 + 0.000012 \times 20)}{1 + 0.000182 \times 20} = 751.44 \text{ mm}$$

Thus height at 0°C = 751.44 mm.

Problem 5 : A barometer having a steel scale reads 755 mm on a day when the temperature is 25°C. If the scale is correctly graduated at 0°C find the true pressure. (Coefficient of linear expansion of steel = 12×10^{-6}; coefficient of cubical expansion of mercury = 182×10^{-6}).

Solution : Actual length of scale at 25°C,

$$H_1 = 755\,(1 + 25 \times 12 \times 10^{-6})$$

Let l_0 be the density of mercury at 0°C, then density of mercury at 25°C,

$$l_{25} = \frac{l}{1 + 25 \times 182 \times 10^{-6}}$$

$\therefore$ True pressure of 25°C = $755\,(1 + 25 \times 12 \times 10^{-6}) l_{25} \cdot g$

or
$$H_0 l_0 g = \frac{755\,(1 + 25 \times 12 \times 10^{-6})}{(1 + 25 \times 182 \times 10^{-6})}$$

$\therefore$ True Barometric Height

$$H_0 = 755\,(1 - 25\,(182 - 12) \times 10^{-6}] \text{ mm}$$

$$= (755 \times 0.99575) \text{ mm}$$

$$= 751.79 \text{ mm.}$$

Problem 6 : A glass bottle, volume 50 cc, at 0°C, is filled with paraffin at 15°C, what is the mass of the paraffin? The density of paraffin at 0°C is 0.82 g per cc; the coefficient of real expansion of paraffin for the range 0° to 15°C is 0.0009 and coefficient of linear expansion of glass is 0.000009 both per °C.

Solution : Coefficient of cubical expansion of glass = 3 × 0.000009 = 0.000027.

$\therefore$ Volume of bottle at 15°C = 50 (1 + 0.0000027 × 15)

$$\text{Density of paraffin at } 15°C = \frac{d_0}{1+\gamma t} = \frac{0.82}{1+0.0009\times 15}$$

$\therefore$ Mass of paraffin at 15°C = Volume × density

$$= 50(1 + 0.000027 \times 15) \times \frac{0.82}{1 + 0.0009 \times 15}$$

$$= 50.02 \times \frac{0.82}{1.0135}$$

$$= 40.47 \text{ g.}$$

Problem 7 : A weight thermometer of fused silica holds 5.00 g of a liquid at 25°C; when heated to 100°C, 0.400g, overflows. Assuming that the liquid expands uniformly through out the range and that the effect of the expansion of the silica is negligilbe, how much more liquid will overflow when the temperature is raised to 200°C.

Solution : The apparent coefficient of expansion of the liquid

$$= \frac{\text{mass expelled}}{\text{mass left} \times \text{rise in temp.}}$$

$$= \frac{0.4}{4.6\times 75} = 0.00116\,°C$$

Suppose m gm of the liquid overflows when the temperature rises from 100°C to 200°C. Then (4.6 – m) gas will be the mass left in the vessel.

$$\text{But apparent coefficient} = \frac{\text{mass expelled}}{\text{mass left} \times \text{rise in temp}}$$

$$\therefore 0.00116 = \frac{m}{(4.6-m)\times 100}$$

$$\therefore 0.00116 \times 100\,(4.6 - m) = m$$

or 1.116 m = 0.534

or m = 0.478 g.

Problem 8 : A loaded glass bulb weighs 156.25 g in air, 57.50 g when immersed in a liquid at 15°C and 58.57g when immersed at 52°C. Calculate the mean coefficient of real expansion of the liquid between 15°C and 52°C. (Coefficient of linear expansion of glass = 0.000009 per degree C).

Solution : Upthrust on bulb at 15°C = (156.25 – 57.50) gm wt

= 98.75 gm. wt.

Upthrust on bulb at 52°C = (156.25 - 58.57) gm. wt.

= 97.68 gm. wt.

But upthrust = at. of liquid displaced

$\therefore\ 98.75 = v_1 d_1$

and $98.68 = v_2 d_2$

where v_1, v_2 are the volumes of glass at 15°C and 52°C respectively and d_1 and d_2 are the corresponding densities of the liquid.

$$\therefore\ \frac{v_1 d_1}{v_2 d_2} = \frac{98.75}{97.68} \quad \ldots.\ \text{(i)}$$

But $v_2 = v_1\ [1 + 0.000027\ (52 - 15)]$

(because cubical expansion of glass = 3 × 0.000009)

If γ is the real coefficient of expansion of the liquid :

or $$\frac{d_1}{d_2} = 1 + \gamma\ (52 - 15)$$

substituting for $\frac{v_1}{v_2}$ and $\frac{d_1}{d_2}$ in eq (i), we get

$$\frac{1 + 37\gamma}{1 + 37 \times 0.000027} = \frac{98.75}{97.68}$$

Solving for γ, $\gamma = 0.00032$ per degree

(using bionomeial theorem)

Problem 9 : A mercury thermometer immersed upto 30° mark in a bath reads 230°C. If the average termperature of the emergent column is 50°C, calculate the correct temperature. (Coefficient of absolute expansion of mercury = 1.82×10^{-4}; coefficient of cubical expansion of glass = 24×10^{-6})

Solution : Coefficient of apparent expansion of mercury

$$= 1.82 \times 10^{-4} - 24 \times 10^{-6} = 158 \times 10^{-6}$$

Let t°C be the true temperature, then

length of the exposed column at t°C = (t – 30)

$$(t - 30) = (230 - 30)[1 + 158\,(t - 50) \times 10^{-6}]$$

$$= 200 + 200 \times 158 \times 10^{-6}\,t - 200 \times 158 \times 50 \times 10^{-6}$$

or $0.9684\,t = 230 - 1.58$

$$= 228.42$$

or $t = 235.8°C.$

4

Expansion of Gases

Problem 1 : It is found that the volume of a certain gas increases in the ratio 1.035 : 1 between 15°C and 25°C. Calculate the absolute zero on the centigrade scale for this gas.

Solution : Suppose the absolute zero is x°C, below 0°C. Then assuming the change of volume takes place at constant pressure, the volume is proportional to the absolute temperature. Hence

$$\frac{x + 25}{x + 15} = \frac{1.035}{1}$$

From which $x = -270.7\ °C$

Problem 2 : The density of argon is 1.60g per litre at 27°C and at a pressure of 75 cm of mercury. What is the mass of argon in an argon filled electric lamp bulb of volume 100 cc. If the pressure inside is 75 cm of mercury when the average temperature of the gas is 120°C.

Solution : The volume of a fixed mass of gas at a constant pressure is proportional to its absolute temperature. Thus the volume V at 27°C of 100 cc of argon at 120°C and 75 cm pressure is given by

$$\frac{V}{100} = \frac{273 + 27}{273 + 120}$$

$$\text{Or} \quad v = \frac{300}{393} \times 100 = 76.3 \text{ cc.}$$

The pressure being constant at 75 cm mercury.

But 1000 cc of argon at 27°C and 75 cm pressure weighs = 1.60g

$\therefore$ 76.3 cc of argon at 27°C and 75 cm pressure $= \dfrac{76.3}{1000} \times 1.60$

$= 0.122$ g.

This is the mass of argon in electric lamp.

Problem 3 : A diving-bell of unform cross-section and 2 meters high is full of air at 70°C when the atmospheric pressure is equal to 77 cm of mercury. It is then sunk with water at 27°C and the water rises 50 cm with in the bell. Assuming that no air escapes as the bell is sunk, find the depth of the bell below the water surface (density of mercury = 13.6 g/cm^3)

Solution : When the level of the water rises 50 cm to A (see Fig.). The volume v_1 of air inside the bell = 150α cc. Where α is the uniform cross-section of the bell in sq. cm.

The pressure p_1 of the air = The pressure at A

$$= 77 \times 13.6\, g + (150 + x)g \text{ dynes/cm}^2$$

Where x is the depth of the top of the bell below the surface of the water.

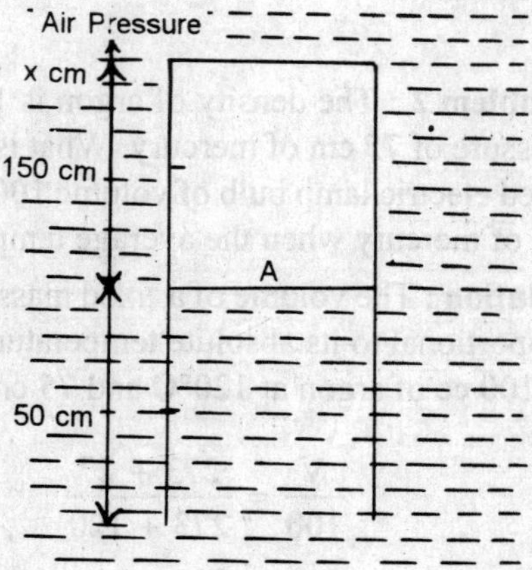

The absolute temperature T_1 of the air = 273 + 27 = 300 K.

Now the origional pressure p_2 of the air in the bell = 77 × 13.6 g dynes/cm^2 and the original volume v_2 of the air in the bell = 200 α cc.

The original absolute temperature = 273 + 70 = 343 K.

Since the mass of the air in the bell is constant

$$\therefore \frac{P_1V_1}{T_1} = \frac{P_2V_2}{T_2}$$

$$\therefore \frac{(77 \times 13.6 + 150 + x)\, g \times 150\alpha}{300} = \frac{77 \times 13.6\, g \times 200\alpha}{343}$$

$$x = 2400 \text{ cm}$$

Problem 4 : Two glass bulbs of volumes 25 and 35 cc are connected by a narrow tube of negligible volume. The apparatus is filled with air at 0°C and 76 cm of mercury pressure and sealed off. If the 25 cc bulb is maintained at 0°C and 35 cc bulb at 100°C. What is the new pressure with in the apparatus?

Solution : At 0°C and 76cm pressure, the total volume of air = 25 + 35 = 60cc. Since the mass of the gas (m) is given by

$$m = \frac{pv}{RT}$$

The total mass of air in two bulbs $= \dfrac{76 \times 13.6 \times 980 \times 60}{(R \times 273)}$

Where R is the gas constant of 1g of air.

The temperature of 35 cc bulb is changed to 100°cc or 373K. Hence the mass of the air in this bulb $= \dfrac{p \times 35}{R \times 373}$

where p is the new pressure in the apparatus.

Similarly the mass of the air now present in 25 cc bulb $= \dfrac{p \times 25}{R \times 273}$, but the total mass of the air is constant. Hence

$$= \frac{p \times 35}{R \times 373} + \frac{p \times 25}{R \times 273} = \frac{76 \times 13.6 \times 980 \times 60}{R \times 273}$$

Dividing through out by $\dfrac{1}{R}$, we get

$$\frac{35p}{373} + \frac{25p}{273} = \frac{76 \times 13.6 \times 980 \times 60}{273}$$

or p = 90.1 cm of mercury

5

First Law of Thermodynamics

Problem 1 : A waterfall is 100 m high. Assuming that all the heat generated stays in water, how much warmer the water at the bottom will become as compared with water at the top?

Solution : Let us take 1 kg of water. When it falls through a height of 100 m, the decrease in its potential energy is

$$\Delta E = 1\text{kg} \times 9.8\ \text{ms}^{-1} \times 100\text{m} = 981\text{J}$$

This decrease in potential energy will first appear as kinetic energy of water as it falls. When it strikes the bottom, this kinetic energy reduces to zero. This now appears as thermal energy. Therefore the thermal energy produced is

$$Q = 981\text{J} = \frac{981}{4.184}\text{Cal} = 234.5\ \text{Cal.}$$

Heat imput required to raise the temperature of 1 kg of water through 1°C is 1 Kcal. Therefore rise in temperature

$$= \frac{234.5}{1000°\text{C}} = 0.23°\text{C}$$

Problem 2 : One mole of perfect gas is allowed to expand from a pressure of atmospheres to a pressure of 1 atmosphere keeping the temperatrue canstant at 27°C (an isothermal process). Draw the PV diagram (or indicator diagram) and from this find out the work done by the gas. Compare it with the theoretical value given by the formula $w = -R_0T\ ln\ p_2/p_1$ were P_1 and P_2 are the initial and final pressure respectively.

Solution : For one mole of an ideal gas

$$PV = R_0T$$

or $$V = \frac{R_0T}{P}$$

$$= 8.314 \text{ TK}^{-1} \text{ mol}^{-1} \times 300\text{K}$$

$$\text{Now } P = 1 \text{ atm } = 1.013 \times 10^6 \text{Nm}^{-2}$$

If we now substitute various values of P we can find out the molar volumes at these pressures. Then we can draw PV-diagram shown below.

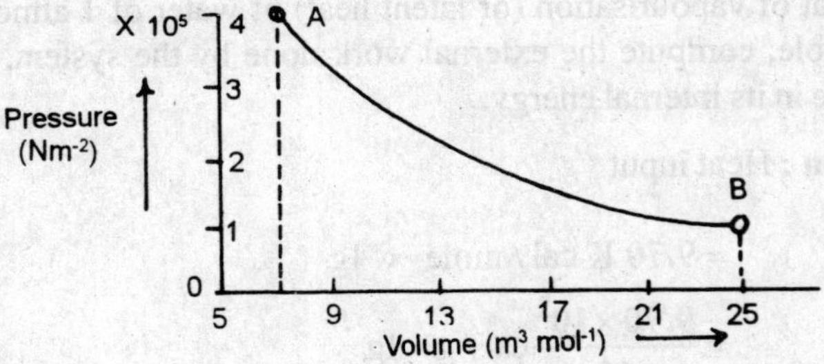

Area enclosed between this curve, the abscissa and the broken lines is 3420J/mole, which is the work done by the gas is going from A to B.

From theoretical formula

$$W = -RT\, ln\frac{P_2}{P_1} = RT\, ln\frac{P_1}{P_2}$$

We get, $$W = 8.314 \times 300 \times 2.303 \times 0.6021$$

$$= 3458 \text{ J/mole}$$

The difference in two results in due to some error involvled in plotting the graph and counting the squares. We take more than half a square as one leave out less than a half.

Problem 3 : A gram molecule of a gas at 27°C expands isothermally until its volume is doubled. Find the work done. R = 8.3 × 10^7 ergs/ gm mol/°C.

Solution : Work done,

$$W = RT\, ln\frac{V_2}{V_1} = 2.303\ \log\frac{V_2}{V_1}$$

$$= 2.303 \times 8.3 \times 10^7 \times 300\ \log\frac{2}{1}$$

$$= 2.303 \times 8.3 \times 10^7 \times 300 \times 0.3010$$

$$= 1.727 \times 10^7 \text{ ergs}$$

Problem 4 : When 1 g of boiling water is converted into steam at a pressure of 1 atmosphere. The steam occupies a volume of 1671 cm^3. The heat of vapourisation (or latent heat) of water of 1 atmo is 9.70 Kcal/mole, compute the external work done by the system, and the increase in its internal energy.

Solution : Heat input

$$= 9.70 \text{ K cal / mole } \times \text{ 1g}$$

$$= \frac{9.70 \times 10^3}{18} \text{cal / g} \times 1\text{g}$$

$$= 539 \text{ cal}$$

$$= 539 \times 4.184\text{J}$$

$$= 2255\text{J}.$$

Now in process of formation of steam the external pressure remain constant at 1 atm. Therefore work done by the system

$$W = \Sigma P\Delta V = P\Sigma\Delta V = P(V_2 - V_1)$$

$$= 1.013 \times 10^5 \text{Nm}^{-2} \times \frac{(1671 - 1)}{10^6} \text{m}^3$$

$$= \frac{1.013 \times 1670}{10} \text{Nm}$$

$$= 169\text{J}$$

Now increase in internal energy ΔE is given by

$$\Delta E = \Delta Q - \Delta W$$

$$\text{or } \Delta E = (E_2 - E_1) = (2255 - 169)\text{J}$$

$$= 2086\text{J}$$

$$= \frac{2086}{4.184}\text{cal}$$

$$= 499\text{cal.}$$

Problem 5: Derive an expression for the work done when a certain mass of a perfect gas undergoes adiabatic changes in pressure and volume, the volume changing from v_1 to v_2 and termperature changing from T_1 to T_2.

Solution : Work done is given by

$$w = \int_{v_1}^{v_2} P dv$$

But the adiabatic process $Pv^{\gamma} = k$ (a constant)

$$\text{or} \quad P = \frac{K}{v^{\gamma}}$$

$$\therefore \quad W = \int_{V_1}^{V_2} \frac{K}{v^{\gamma}} dV = k\left\{\frac{V^{1-\gamma}}{1-\gamma}\right\}_{V_1}^{V_2}$$

$$= \frac{k}{1-\gamma}\left\{V_2^{1-\gamma} - V_1^{1-\gamma}\right\}$$

$$\text{Since} \quad \chi = P_1V_1^{\gamma} = P_2V_2^{\gamma}$$

$$\therefore \quad W = \frac{1}{1-\gamma}\left\{P_2V_2^{\gamma}.V_2^{1-\gamma} - P_1V_1^{\gamma}.V_1^{1-\gamma}\right\}$$

$$= \frac{1}{1-\gamma}\left\{P_2V_2 - P_1V_1\right\} = \frac{1}{1-\gamma}\left\{P_1V_1 - P_2V_2\right\}$$

When the temperature changes from T_1 to T_2, then

$$P_1V_1 = RT_1 \text{ and } P_2V_2 = RT_2$$

$$\therefore \quad W = \frac{1}{\gamma - 1}(T_1 - T_2)$$

Problem 6 : A gram molecule of a gas at 27°C expands adiabatically until its volume is doubled. Calculate the work done. $R = 8.3 \times 10^7$ ergs degree^{-1} mol^{-1}; $\gamma = 1.4$.

Solution : Let V_1, T_1 be initial volume and temperature and V_2, T_2, be the final volume and temperature respectively, then

$$\frac{V_2}{V_1} = 2;\ T_1 = 27 + 273 = 300\,\text{A}$$

$$\text{Now}\quad T_1 V_1^{\gamma-1} = T_2 V_2^{\gamma-1}$$

$$\therefore\ T_2 = T_1\left(\frac{V_1}{V_2}\right)^{\gamma-1} = 300 \times \left(\frac{1}{2}\right)^{0.4}$$

$$= 227.4\,\text{A}.$$

Now work done , $W = \dfrac{R}{\gamma - 1}(T_1 - T_2)$

$$= \frac{8.3 \times 10^7}{0.4}(300 - 227.4)$$

$$= 1507 \times 10^7 \text{ ergs}$$

$$= 1507 \text{ Joules}$$

Problem 7 : Show that the slope of adiabatic is γ times the slope of isothermal under the same canditions of pressure and volume.

Solution : For an isothermal, PV = k (constant)

Differentiating, we get

$$PdV + VdP = 0$$

$$\therefore\ PdV = -VdP$$

$$\text{or}\quad \frac{dP}{dV} = -\frac{P}{V}$$

since $\dfrac{dP}{dV}$ gives the slop, so

$$\text{slope of isothermal} = -\frac{P}{V} \qquad \ldots\text{(i)}$$

For an adiabatic process, we have

$$PV^{\gamma} = \chi(\text{constant})$$

Differentating, we get

$$V^{\gamma} dP + P\gamma\, V^{\gamma-1} dV = 0$$

$$\text{or} \quad VdP + P\gamma\, dV = 0$$

$$\frac{dP}{dV} = -\gamma . \frac{P}{V}$$

Since $\frac{dP}{dV}$ gives the slope

$$\therefore \quad \text{Slope of adiabatic} = -\gamma . \frac{P}{v} \quad .... \text{(ii)}$$

Comparing (i) and (ii), we get

$$\frac{\text{slope of adiabatic}}{\text{slope of isothermal}} = \frac{-\gamma \cdot \frac{P}{v}}{-\frac{P}{v}}$$

$$= \gamma$$

Problem 8 : In an experiment to find the specific heat of tin, a 200.0 g piece of tin at 98.0°C is put into a copper calorimeter cantaining 100 g of water at 20.0°C. The temperature of the mixture becomes 27.1°C. Find out the specific heat of tin. The mass of calorimeter is 102.0 g and the specific heat of copper is 0.094 Kcal/Kg K. The specific heat of water is 1 Kcal/Kg K.

Solution : Let the specific heat of tin be S

$$\text{Heat lost by tin piece} = \frac{200.0}{1000} \text{Kg} \times 5 \times (98 - 27.1) \text{ K}$$

$$= 0.2 \times 5 \times 70.9 \text{Kg K}$$

$$\text{Heat gained by water} = \frac{100.0}{1000} \text{kg} \times 1 \text{kcal / kg} - \text{k} \times (27.1 - 20.0) \text{k}$$

$$= 0.1 \times 7.1 \text{kcal.}$$

Now Heat gained by calorimeter

$$= \frac{102.0}{1000} \text{kg} \times 0.094 \text{ Kcal / Kg} - \text{K} \times (27 - 20)\text{K}$$

$$= 0.102 \times 0.094 \times 7.1\text{k cal.}$$

Now for the composite system of tin and water, there is no net heat loss or heat gain, therefore we have.

$$= 0.2 \times 5 \times 70.9 = 0.1 \times 7.1 + 0.102 \times 0.094 \times 7.1$$

$$\text{or} \quad 14.18 \text{ S} = (0.71 + 0.068)\text{Kcal / Kg} - \text{K}$$

$$\text{or} \quad \text{S} = \frac{0.778}{14.18} \text{Kcal / Kg} - \text{K}$$

$$= 0.055\text{Kcal / Kg} - \text{K}$$

∴ Specific heat of tin = 0.055 kcal/kg–k.

Problem 9 : Specific heat of argon at constant pressure is 0.1253 cal/g–°C at 15°C and one atmosphere pressure. Further $\frac{c_p}{c_v} = \gamma = 1.668$. Find the volume of R, assuming argon to behave as a perfect gas.

Solution :

Given $\quad C_p = 0.1253\text{cal / g–°C}$

$$= 0.1253 \times 39.95 = 5.005\text{cal / mole–°C}$$

$$C_v = \frac{5.005}{1.668} = 3.000\text{cal / mole–°C}$$

$$R = C_p - C_v$$

Thus

$$= 5.005 - 3.000$$

$$= 2.005\text{cal / mole–°C}$$

If we compare it with standard value of R (i.e. 1.987 cal/mole–°C) we find the difference is quite small and so it can be concluded that at moderate pressure argon behaves like a perfect gas.

Problem 10 : Calculate the efficiency of a carnot's cycle working between the temperatures 125°C and 15°C

Solution : Efficiency, $\eta = \frac{T_1 - T_2}{T_1} \times 100$

Given : Temperature of source, $T_1 = 125°C = 125 + 273 = 398$ A

Temperature of sink, $T_2 = 15°C = 15 + 273 = 288A$

$$\therefore \quad \eta = \frac{398 - 288}{398} \times 100 = 27.6\%$$

Problem 11 : Derive a mathematical expression for the efficiency of an internal combustion engine (The **Otto engine**)

Solution : To derive the expression, let us consider 1 gram molecule of the working substance. The heat Q_1 supplied to it by combution is given by

$$Q_1 = C_v (T_3 - T_2)$$

Where C_v in specific heat at constant volume.

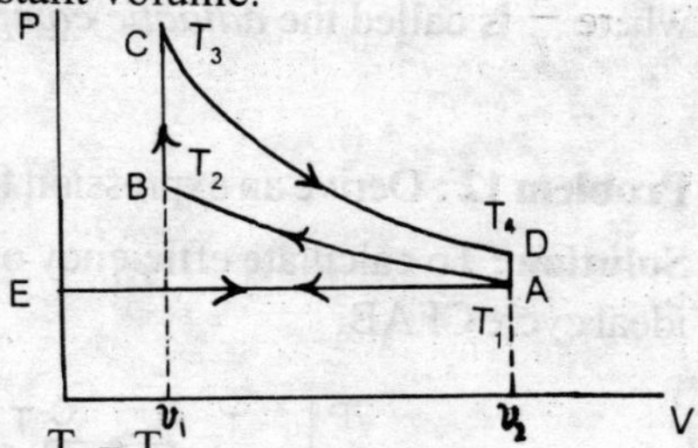

The heat Q_2 rejected is given by

$$Q_2 = C_v (T_u - T_1)$$

Hence efficiency, $\eta = 1 - \frac{Q_2}{Q_1} = 1 - \frac{T_4 - T_1}{T_3 - T_2}$ . . . .(i)

Since ABCD is an adiabatic,

$$T_1 V_2^{\gamma-1} = T_2 V_1^{\gamma-1}$$

where γ is the ratio of specific heats of the working substance

$$\therefore \quad \frac{T_1}{T_2} = \left(\frac{V_1}{V_2}\right)^{\gamma-1} \quad . . . .(ii)$$

Similiarly since CD is an adiabatic

$$T_3 . V_1^{\gamma-1} = T_4 V_2^{\gamma-1}$$

$$or \quad \frac{T_4}{T_3} = \left(\frac{V_1}{V_2}\right)^{\gamma-1} \qquad \text{....(iii)}$$

$$\therefore \frac{T_4}{T_3} = \frac{T_1}{T_2} = \frac{T_4 - T_1}{T_3 - T_2} = \left(\frac{V_1}{V_2}\right)^{\gamma-1}$$

using his result in (i), we get

$$\eta = 1 - \left(\frac{V_1}{V_2}\right)^{\gamma-1}$$

Since $\frac{V_2}{V_1}$ is the adiabatic expression ratio, ℓ,

$$\eta = 1 - \left(\frac{1}{\ell}\right)^{\gamma-1}$$

where $\frac{1}{\ell}$ is called the *adiaatic compression ratio.*

Problem 12 : Derive an expression for the efficiency of a diesel engine.

Solution : To calculate efficiency of such an engine let us consider an ideal cycle CFAB.

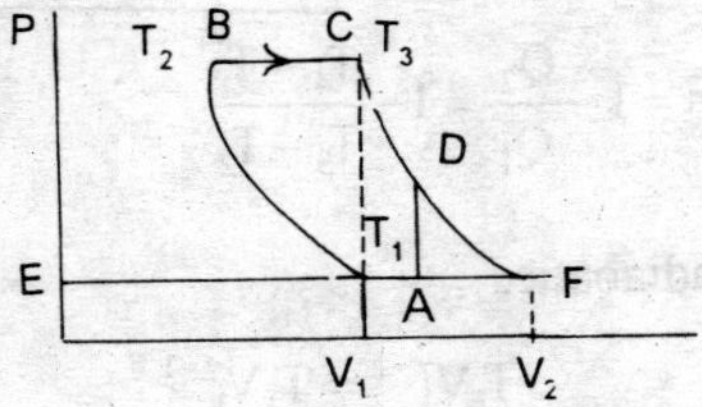

Fig.

Let the specific volumes and pressures of the working substance at C and F be V_1 and V_2 and P_2 and P_1 respectively. Let T_1, T_2, T_3, T_4 be the temperatures at A, B, C and F respectively.

The heat supplied by combustion of fuel (1 gram molecule) along BC is given by

$$Q_1 = C_p (T_1 - T_4)$$

where C_p is the specific heat at constant pressure.
The heat rejected outside along FA is given by

$$Q_2 = C_p (T_4 - T_1)$$

Thus efficiency, $\eta = 1 - \frac{Q_1}{Q_2} = 1 - \frac{T_4 - T_1}{T_3 - T_2}$ . . . .(i)

Since A and B lie on the same adiabatic

$$\therefore P_1^{1-\gamma}.T_1^{\gamma} = P_2^{1-\gamma}.T_2^{\gamma}$$

or $\frac{T_1}{T_2} = \left(\frac{P_2}{P_1}\right)^{\frac{1-\gamma}{\gamma}}$ . . . .(ii)

Similary since C and F lie on the same adiabatic

$$P_2^{1-\gamma}.T_3^{\gamma} = P_1^{1-\gamma}.T_4^{\gamma}$$

or $\frac{T_4}{T_3} = \left(\frac{P_2}{P_1}\right)^{\frac{1-\gamma}{\gamma}}$ . . . .(iii)

Comparing (ii) and (iii) we have

$$\frac{T_1}{T_2} = \frac{T_4}{T_3} = \frac{T_4 - T_1}{T_3 - T_2} = \left(\frac{P_2}{P_1}\right)^{\frac{1-\gamma}{\gamma}}$$

Again C and F lie on the same adiabatic.

$$P_2 . V_1^\gamma = P_1 . V_2^\gamma$$

or $$\frac{P_2}{P_1} = \left(\frac{V_2}{V_1}\right)^\gamma$$

$$\frac{T_4 - T_1}{T_3 - T_2} = \left(\frac{P_2}{P_1}\right)^{\frac{1-\gamma}{\gamma}} = \left(\frac{V_2}{V_1}\right)^\gamma$$

Substituting this in (i) we get

$$\eta = 1 - (\ell)^{\gamma - 1} = 1 - \left(\frac{1}{\ell}\right)^{\gamma - 1} \quad \text{....(iv)}$$

This relation is same as for a carnot cycle or an otto cycle.

However, the above relation is based on the assumption that the cycle is ideal one i.e. heat would be rejected at constant pressure along FA. But actually the exhaust value is opened at D and heat is rejected at constant volume along DA. Hence

$$Q_2 = C_v (T_4 - T_1)$$

Taking temperatures at D and F to be practically the same

For the actual diesel cycle

$$\eta = 1 - \frac{C_v(T_4 - T_1)}{C_p(T_3 - T_2)}$$

$$= 1 - \frac{1}{\gamma}\left(\frac{1}{\ell}\right)^{\gamma - 1}$$

The efficiency for the ideal cycle can be calculated and is found to be about 63%. The efficiency for actual cycle is, therefore, some what lower, about 55%.

6

Second Law of Thermodynamics

Problem 1 : Taking the boiling point of water as 100°C and normal latent beat of vapourisation 540 cal. Calculate the elevation of the boiling point for an increase of pressure of 1 atmosphere. One gram of steam occupies is a volume of 1677 cc.

Solution :

Specific volume of steam, $V_2 = 1677$ cc.

Specific volume of water, $V_1 = 1$ cc

Normal boiling point $T = 100°C = 373A$

Increase of pressure $dp = 1$ atmosphere

$= 76 \times 13.6 \times 980$ dynes

Latent heat of vapourisation, $L = 540$ calories

$= 540 \times 4.2 \times 10^7$ ergs

Now
$$\frac{dP}{dT} = \frac{L}{T(V_2 - V_1)}$$

$$\therefore \frac{76 \times 13.6 \times 980}{dT} = \frac{540 \times 4.2 \times 10^7}{373(1677 - 1)}$$

or $dT = 27.92°C$

Problem 2 : The normal melting point of ice is 0°C and the latent heat of fusion 80 calories. If 1 cc. of water on solidification has a volume of 1.091 cc. Calculate the depression of freezing point for a increase of pressure of 1 atmosphere.

Solution :

Specific volume of water, $V_2 = 1$ cc.

Speicific volume of ice, $V_1 = 1.091$ cc

Normal melting point $= 0°C = 273K$

Increase of perssure, $dP = 1$ atmopshere

$= 76 \times 13.6 \times 980$ dynes

Latent heat of fusion, $L = 80$ calories

$= 80 \times 4.2 \times 10^7$ ergs

Now
$$\frac{dP}{dT} = \frac{L}{T(V_2 - V_1)}$$

$$\therefore \frac{76 \times 13.6 \times 980}{dT} = \frac{80 \times 4.2 \times 10^7}{273(1 - 1.091)}$$

$$dT = -0.0074°C \cdot$$

7

Second Law of Thermodynamics

Problem 1 : A 42 kilowatt engine is operating between 227°C, and 177°C, calculate :

(i) The efficiency of the engine

(ii) The amount of heat absorbed and the amount of heat rejected per second.

(Rohilkhand 1990)

Solution :

(i) Efficieny, $\eta = 1 - \frac{T_2}{T_1}$

$$= 1 - \frac{450}{500} \left[\because T_2 = 177 + 273 = 450, T_1 = 227 + 273 = 500\right]$$

$$= 1 – 0.9 = 0.1 \text{ or } 10\%$$

(ii) If Q_1 and Q_2 are the quantities of heat absorbed and heat rejected respectively, then

$$\eta = 1 - \frac{Q_2}{Q_1} = \frac{Q_1 - Q_2}{Q_1} = \frac{W}{Q_1}$$

Given $W = 42 \times 10^3 \text{ watt} = 42 \times 10^3 \text{ J/S}$

$$Q_1 = \frac{W}{\eta} = \frac{42 \times 10^3}{0.1} = 4.2 \times 10^5 \text{ J/S}$$

and $Q_2 = Q_1 - w$

$$= 4.2 \times 10^5 - 42 \times 10^3$$

$$= 4.2 \times 10^5 - 0.42 \times 10^5$$

$$= 3.78 \times 10^5 \text{J / S}$$

Problem 2 : A carnot engine has the same efficiency between 1000K and 500K and between xK and 1000K (his being the temperature of the sink in this case). Calculate x.

[Agra 1971S]

Solution : The efficiency of carnot engine is given by

$$\eta = 1 - \frac{T_2}{T_1}$$

In the first case,

$$\eta = 1 - \frac{500}{1000}$$

$$= 1 - 0.5 = 0.5 \text{ or } 50\%.$$

In the second case

$$\eta_2 = 1 - \frac{1000}{x}$$

Since $\eta_1 = \eta_2$

$$0.5 = 1 - \frac{1000}{x}$$

or $$\frac{1000}{x} = 1 - .5$$

or $$x = 2000\text{K}$$

Problem 3 : A carnot's engine takes in a thousand kilo calories of heat from a reservior at 627C° and exhaust it to sink at 27°C. What is its efficiency?

How much work does it perform? Express it in (i) ergs (ii) kilowatt hours (iii) electron volts.

[Vikram 1977]

Solution :

$$\eta = 1 - \frac{Q_2}{Q_1} = 1 - \frac{T_2}{T_1}$$

$$\therefore \ \eta = 1 - \frac{300}{900} \qquad \left[\begin{array}{l} \therefore T_1 = 627 + 273 = 900\text{K}, \\ T_2 = 27 + 273 = 300\text{K} \end{array}\right]$$

$$= \frac{2}{3} \ \text{or} \ 66.6\%$$

(i) Work done, $W = \eta \times Q_1 = \frac{2}{3} \times 10^6 \text{colories}$

$$\left[\therefore \ 1000\text{k cal} = 10^6 \text{cal}\right]$$

$$= \frac{2}{3} \times 10^6 \times 4.2 \times 10^7 \text{ ergs}$$

$$= 2.8 \times 10^{13} \text{ ergs or } 2.8 \times 10^6 \text{ J}.$$

(ii) 1 kilowatt hour = 1 kilo watt × 1 hour
= 1000 watt × 3600 Sec
= 1000 J/s × 3600s
$= 3.6 \times 10^6$ J

Work done = 2.8×10^6 J

$$= \frac{2.8 \times 10^6 \text{ J}}{3.6 \times 10^6} \text{kilowatt} - \text{hour} \quad \left[1\text{kwt} = \text{hr} = 3.6 \times 10^6 \text{J}\right]$$

$$\frac{7}{9} \text{kilowatt} - \text{hour}$$

(iii) As 1 electron volt $= 1.6 \times 10^{-12}$ ergs

$$W = \frac{2.8 \times 10^{13}}{1.6 \times 10^{-12}} \text{eV}$$

$$= 1.7 \times 10^{25} \text{ eV}.$$

Problem 4 : The efficiency of a Carnot's cycle is $\frac{1}{6}$. If on reducing the temperature of sink by 65°C the efficiency becomes 1/3, find the initial and final temperatures between which the cycle is working.

[Allahabad 1989]

Solution : Let T_1 and T_2 be the initial and final temperatures (i.e. the temperature of source and sink) respectively. Then

In first case, $\eta_1 = 1 - \frac{T_2}{T_1}$

$$\text{or} \quad \frac{1}{6} = 1 - \frac{T_2}{T_1} \qquad \ldots(i)$$

In second case, $\eta_2 = 1 - \frac{T_2 - 65}{T_1}$

$$\text{or} \quad \frac{1}{3} = 1 - \frac{T_2 - 65}{T_1} \qquad \ldots(ii)$$

From (i) and (ii), we get,

$$\frac{65}{T_1} = \frac{1}{6} \qquad \text{[By substracting (i) from (ii)]}$$

$$\text{or } T_1 = 65 \times 6 = 390\text{K. or } 117°\text{C}$$

$$\frac{1}{6} = 1 - \frac{T_2}{390} \qquad \text{[From (i)]}$$

$$\text{or} \quad \frac{T_2}{390} = 1 - \frac{1}{6} = \frac{5}{6}$$

$$\text{or} \quad T_2 = \frac{390 \times 5}{6} = 325\text{k or } 52°\text{C}$$

Problem 5 : A reversible engine takes in heat from a reservior of heat at 527°C and gives out heat to the sink at 127°C. How many calories per second must it take from the reservoir in order to produce useful mechanical work at the rate of 750 watt.? [Agra 1974]

Solution :

$$\eta = 1 - \frac{T_2}{T_1}$$

$$= 1 - \frac{400}{800} \qquad \begin{bmatrix} T_1 = 127 + 273 = 400\text{K}, \\ T_2 = 527 + 273 = 800\text{K} \end{bmatrix}$$

$$= 1 - \frac{1}{2}$$

$$= 0.5$$

Work, $W = 750$ Watt (given)

$$= 750 \text{ J/s}$$

$$= \frac{750}{4.2} \text{cal/s}$$

Since $\eta = \dfrac{W}{Q_1}$

$$Q_1 = \frac{W}{\eta} = \frac{750}{4.2} \times \frac{1}{0.5}$$

$$= 357.17 \text{ Cal/Sec.}$$

Problem 6 : Define or refute the following statement from a theoretical view point it is claimed that with temperature of surroundings 300 K, 1200 kcal of heat available from a reserviour at 900 K is more useful than 15000 kcal of heat available at 600 K.

[Agra 1975]

Solution :

In the first case $\eta_1 = 1 - \dfrac{300}{900} = 1 - \dfrac{1}{3} = \dfrac{2}{3}$

In second case $\eta_2 = 1 - \dfrac{300}{600} = 1 - \dfrac{1}{2} = \dfrac{1}{2}$

Since efficiency, $\eta = \dfrac{\text{work}}{\text{heat}}$

$$\eta_1 = \frac{2}{3} = \frac{W_1}{12000} \text{ or } \qquad W_1 = 8000\text{kcal}$$

$$\text{and } \eta_2 = \frac{1}{2} = \frac{W}{15000} \text{ or } W_2 = 7500\text{kcal}$$

So, $W_1 > W_2$, so the statement is to be defended.

Problem 7 : A carnot engine whose low temperature reservoir at 7°C has an efficiency 5%. It is deserved to increase the efficiency to 70%. By how much degrees should the temperature of the high temperature reservoir be increased?

[Delhi 1971]

Solution :

$$\text{We know, } \eta = 1 - \frac{T_2}{T_1}$$

$$0.5 = 1 - \frac{280}{T_1} \qquad [T_2 = 7 + 273 = 280\text{K}]$$

$$\text{or} \quad T_1 = 560\text{K}.$$

$$\text{Again} \quad \eta^1 = 70\% \text{ or } 0.7$$

$$\text{Then} \quad 0.7 = 1 - \frac{280}{T_1^1}$$

$$\text{or} \quad T_1^{\ 1} = 933.3 \text{ K}$$

So the required increase in temperature of high temperature reservoir is (933.3 – 560) = 373.3K.

Problem 8 : Two carnot engines A and B are operated in series. The first one A receives heat at 900 K and rejects to a reservoir at temperature TK. The second engine B receives the heat rejected by the first engine and in turn rejects to a heat reservoir at 400 K. Calculate temperature T for the situation when :

(i) the work outputs of the two engines are equal

(ii) The efficiencies by the two engines are equal.

[Lucknow 1979]

Solution :

$$\eta = \frac{\text{work}}{\text{Heat}} = \frac{W}{Q_1} = \frac{Q_1 - Q_2}{Q_1}$$

Also $$\eta = \frac{T_1 - T_2}{T_1}$$

Thus for Engine 'A', $$\eta_1 = \frac{W}{Q_1} = \frac{900 - T}{900}$$

for Engine 'B', $$\eta_2 = \frac{W^1}{Q_1^1} = \frac{T - 400}{T}$$

(i) *When work output of engines 'A' and 'B' are equal,*

then $W = W'$

or $900 - T = T - 400$

or $2T = 1300$

or $T = 650K$.

(ii) *When the efficiencies of engines 'A' and 'B' are equal*

$$\eta_1 = \eta_2$$

or $$= \frac{900 - T}{900} = \frac{T - 400}{T}$$

or $900\,T - T^2 = 900T - 360000$

or $T^2 = 360000$

or $T = 600$ K.

Problem 9 : Two carnot engines 'A' and 'B' are operated in series. First engine 'A' absorbs heat at 1100 K and rejects it to the sink at TK, Second engine 'B' absorbs half amount of heat rejected by first engine and rejects heat to the sink at 200K. If the work performed by both the engines are equal, calculate the temperature T.

[Rohilkhand 1984]

Solution : If engine 'A' absorbs heat 1100 K and rejects heat at TK, then efficiency, η_A is given by,

$$\eta_A = 1 - \frac{T}{1100} = 1 - \frac{Q_2}{Q_1} \text{ or } \frac{Q_2}{Q_1} = \frac{T}{1100}$$

where Q_1 is the heat absorbed and Q_1 is the heat rejected by engine 'A'

The engine 'B' absorbs heat at TK and rejects heat at 200K, then its efficiency, η_B is

$$\eta_B = 1 - \frac{200}{T} = 1 - \frac{Q_3}{Q_2/2} \text{ or } \frac{2Q_3}{Q_2} = \frac{200}{T} \text{ or } \frac{Q_3}{Q_2} = \frac{100}{T}$$

where Q_3 is the heat rejected by engine 'B' and $\frac{Q_2}{2}$ is the heat absorbed by it.

[Heat absorbed $= \frac{1}{2} \times$ heat rejected by engine A.

$$= \frac{1}{2} \times Q_2 = \frac{Q_2}{2}$$

Since the work performed by two engines is equal i.e.

$$W_A = W_{13}$$

$$\therefore \; Q_1 - Q_2 = \frac{Q_2}{2} - Q_3$$

or $\quad Q_1 - \frac{3}{2} Q_2 + Q_3 = 0$

or $\quad 1 - \frac{3}{2}\frac{Q_2}{Q_1} + \frac{Q_3}{Q_1} = 0$

or $\quad 1 - \frac{3}{2}\frac{Q_2}{Q_1} + \frac{Q_3}{Q_1} \cdot \frac{Q_2}{Q_1} = 0$

or $\quad 1 - \frac{3}{2} \cdot \frac{T}{1100} + \frac{100}{T} \times \frac{T}{1100} = 0$

or $\quad \frac{3T}{2200} = 1 + \frac{1}{11}$

$$\text{or} \quad \frac{3T}{2200} = \frac{12}{11}$$

$$\text{or} \quad 33T = 26400$$

$$\text{or} \quad T = \frac{26400}{33}$$

$$= 800\text{K}.$$

Problem 10 : A refrigerator has to transfer an average of 200 J of heat per second from temperature –10°C to + 27°C. Calculate the average power consumed, assuming ideal reversible cycle and no other losses. [Agra 1976]

Solution : We know

$$\eta = 1 - \frac{T_2}{T_1} = 1 - \frac{Q_2}{Q_1} \quad \text{[For ideal reversible cycle]}$$

$$\text{or} \quad \frac{T_2}{T_1} = \frac{Q_2}{Q_1}$$

$$\text{or} \quad Q_1 = Q_2 \cdot \frac{T_1}{T_2}$$

Given : $Q_2 = 200$ J/S ; $T_1 = 27°C = 300$ K

$T_2 = -10°C = 263$ K.

Putting these values in above equation,

$$Q_1 = 200 \times \frac{300}{263} = 228.1 \text{J / S}$$

Hence average power consumed $= W = Q_1 - Q_2$

$= 228.1 - 200$

$= 28.1$ J/S

$= 28.1$ Watt

$$\text{or} = \frac{28.1}{746} \text{HP}$$

$= 0.038$ HP

Problem 11 : A carnot refrigerator takes heat from water at 0°C and discards it to the room at 27°C. If 100 kg of water at 0°C are to be changed to ice at 0°C, what is the required work in joules? Latent heat of ice = 3.4×10^5 J/kg [Delhi 1974]

Solution : We know

$$\frac{Q_1}{Q_2} = \frac{T_1}{T_2} \quad \text{[For an ideal reversible cycle]}$$

or $$Q_1 = Q_2 \cdot \frac{T_1}{T_2}$$

Given : $Q_2 = 100 \times 3.4 \times 10^5$ J; $T_1 = 300$ K; $T_2 = 273$ K.

Putting these values in the above equation,

$$Q_1 = \frac{3.4 \times 10^5 \times 100 \times 300}{273}$$

$$= 3.736 \times 10^7 \text{ J}$$

Hence work done on refrigerator,

$$W = Q_1 - Q_2$$

$$= (3.336 \times 10^7 - 3.4 \times 10^7) \text{ J}$$

$$= 3.336 \times 10^7 \text{J}$$

Problem 12 : It was claimed by an inventor that he has developed an engine that works between 600 K and 300 K and is capable of having efficiency of 60%. Is his claim valid, if not, why?

Solution : The efficiency of carnot engine working between 600 K and 300 K is given by; $n = 1 - \frac{300}{600} = 0.5 = 50\%$.

The claim is not valid because no engine can be more efficient than carnot engine.

8

Production of Very Low Temperatures

Problem 1 : A refrigerator works under reversible cycle between the temperatures 450 K and 600 K. Calculate (i) the thermal efficiency (ii) the coefficient of performance.

Solution :

(i) Thermal efficiency,

$$\eta = 1 - \frac{T_2}{T_1} = 1 - \frac{450}{600} = 0.25 \text{ or } 25\%$$

(ii) The coefficient of performance, p,

$$\beta = \frac{\text{Amount of heat absorbed from cold reservoir}}{\text{Work done in running the machinery}}$$

$$= \frac{Q_2}{W} = \frac{Q_2}{Q_1 - Q_2} = \frac{T_2}{T_1 - T_2}$$

Also $\beta = \dfrac{1-\eta}{\eta}$

Given : $T_2 = 450$ K; $T_1 = 600$ K

$$\beta = \frac{450}{600 - 450} = 3$$

Problem 2 : For a refrigerator heat works under reversible cycle between the temperatures 450 K and 600 K. Calculate refrigerating effect of 100 watts if the refrigerator works at (i) 100% coefficient (ii) 60% coefficient.

Solution : For such an engine, the coefficient of performance

$$\beta_{max} = \frac{T_2}{T_1 - T_2} = \frac{450}{600-450} = 3$$

(i) For 100% coefficient $B_{actual} = B_{max} = 3$

But $\beta = \frac{Q_2}{W}$

So the work done (power needed) for refrigerating effect of 100 watt

$$W = \frac{Q_2}{\beta_{actual}} = \frac{100}{3} = 33.3\text{watt}$$

(ii) For 60% coefficient

$$\beta_{actual} = \frac{60}{100} \times 3 = 1.8$$

$$W = \frac{Q_2}{\beta_{actual}} = \frac{100}{1.8} = 55.5\text{watt}.$$

Problem 3 : Calculate the least amount of work, expressed in kilowatt hours required to convert 100 g of water at 30°C into ice at –20°C using a reservoir of cooling water at 30°C. Assume heat the pressure remains constant throughout the process. Given C_p of ice = 0.50 kcal/kg °C.

Solution :

Given T_1 = 30°C = 303 K; T_2 = – 20°C = 253 K.

m = 100 g = 0.1 kg, C_p = 0.5 kcal/kg °C.

Quantity of heat taken up from the reservoir,

$$Q_2 = 0.1 \times 30 + 0.1 \times \frac{3.4 \times 10^5}{4.2 \times 10^3} + 0.1 \times 0.5 \times 20\text{Kcal}$$

$$= (3.0 + 8.10 + 10)\ \text{Kcal}$$

$$= 21.1\ \text{Kcal}$$

For ideal refrigerator, $\beta = \frac{Q_2}{W} = \frac{T_2}{T_1 - T_2}$

So work required = $Q_2 \cdot \dfrac{T_1 - T_2}{T_2}$

$$= 21.1 \times \frac{303 - 253}{253}$$

$$= 21.1 \times \frac{50}{253} \text{ kcal}$$

$$= 21.1 \times \frac{50}{253} \times 4.2 \times 10^3 \text{ J}$$

$$= 21.1 \times \frac{50}{253} \times 4.2 \times 10^3 \text{ watt} - \text{sec}$$

$$= 21.1 \times \frac{50}{253} \times \frac{4.2 \times 10^3}{3600 \times 10^3} \text{ K.W.H.}$$

$$= 4.89 \times 10^{-3} \text{ K.W.H.}$$

Problem 4 : In a vapour compression machine ammonia is used as a refrigerant. The refrigerator is maintained at – 16°C and the temperature of coolant is 32°C. From Mollieri liquid ammonia chart the total heat contents of ammonia (enthalpies of ammonia) at points A, B and C are 77.2, 340.7 and 396.3 kcal per kg of NH_3 respectively. Calculate the efficiency of performance of the machine and the relative efficiency of machine taking the efficiency of the ideal carnot's refrigerator working between the same temperature to be the maximum.

Solution :

Given : $H_A = 77.2$ kcal per kg of NH_3

$H_B = 340.7$ kcal per kg of NH_3

$H_C = 396.3$ kcal per kg of NH_3

∴ The coefficient of performance of the machine

$$\beta = \frac{H_B - H_A}{H_C - H_B}$$

$$= \frac{340.7 - 72.2}{396.3 - 340.7}$$

$$= \frac{263.5}{55.6} = 4.74$$

The coefficient of performance of an ideal carnot's refrigerator is given by

$$\beta_{max} = \frac{T_2}{T_1 - T_2} = \frac{273 - 16}{(273 + 32) - (273 - 16)}$$

$$= \frac{257}{305 - 257} = \frac{257}{48}$$

$$= 5.35$$

$$\therefore \text{Relative efficiency of machine} = \frac{4.74}{5.35} = 0.888$$

$$= 88.8\%$$

Problem 5 : Calculate the coefficient of performance of an adiabatic expansion refrigerator using oxygen at compression ratio 5, γ for oxygen = 1.40 and $5^{2/7} = 1.58$

[Rohilkhand 1981S]

Solution : Coefficient of Performance of a gas

$$\text{expansion machine} = \frac{1}{\left[\left(\frac{P_2}{P_1}\right)^{\frac{\gamma-1}{\gamma}} - 1\right]}$$

Given : $\frac{P_2}{P_1} = 5; \gamma = 1.4$

$$\beta = \frac{1}{(5)^{\frac{1.4-1}{1.4}} - 1}$$

$$= \frac{1}{5^{2/7} - 1}$$

$$= \frac{1}{1.58 - 1} = \frac{1}{0.58} = 1.72$$

Problem 6 : The compressor of an air expansion machine intakes air at 2 atmosphere and releases at 10 atmospheres pressure. Calculate the coefficient of performance of machine. If the power necessary to drive the machine is 2.0 H.P. Find the amount of heat removed per minute. γ for air = 1.4.

Solution : The coefficient of performance is given by

$$\beta = \frac{Q_2}{W} = \frac{1}{\left(\frac{P_2}{P_1}\right)^{\frac{\gamma-1}{\gamma}} - 1}$$

Given : $P_2 = 6$ atmospheres, $P_1 = 2$ atmospheres

$r = 1.4$, $W = 2.0$ HP $= 2.0 \times 746$ J/S

Substituting the values in equation we get

$$\beta = \frac{1}{\left(\frac{10}{2}\right)^{\frac{1.4-1}{1.4}} - 1}$$

$$= \frac{1}{5^{\frac{.4}{1.4}} - 1} = 1.72$$

Heat removed per minute

$$Q = \beta . W.$$

$$= 1.72 \times 2 \times 746 \times 60 J = \text{min}^{-1}$$

$$= 154 \times 10^3 J = \text{min}^{-1}$$

$$= \frac{154 \times 10^3}{4.2} \text{cal/min}$$

$$= \frac{154}{4.2} \text{kcal/min}$$

$$= 36.6 \text{kcal/min}$$

Problem 7 : Calculate the cooling produced by adiabatic demagnetisation of paramagnetic salt as the field as is reduced from 5000 oersted to zero at initial temperature of 2K. Given curie constant per gram mole per cc = 0.06 C.G.S. units and $C_H = 0.42$ J gm^{-1} deg^{-1}.

[Rajasthan 1984]

Solution : The final temperature is given by

$$T_2 = T_1 \left[1 - \frac{K}{C_H}\left(\frac{H}{T_1}\right)^2\right]^{\frac{1}{2}}$$

Given : $T_1 = 2K$; $C_H = 0.42\ Jgm^{-1}\ deg^{-1} = 0.42 \times 10^7$ ergs $gm^{-1}\ deg^{-1}$
Curie constant, $K = 0.06$

$$\therefore\ T_2 = 2\left[1 - \frac{0.06}{0.42 \times 10^7}\left(\frac{500}{2}\right)^2\right]^{\frac{1}{2}}$$

$$= 1.82\ K.$$

$\therefore$ Cooling produced $= T_1 - T_2 = 2 - 1.82 = 0.18$ K.

9

The Mechanical Equivalent of Heat

Problem 1 : A motor car weighing 2 tons is brought to rest by its well adjusted four-wheel brakes from a speed of 90 m.p.h. If the brake mechanism on each wheel weights 16 lb. Calculate the greatest possible rise in temperature.

(Specific heat of brake material = 0.140; J = 1400 ft.-lb.°C)

Solution :

Work done by brakes in bringing car to rest $= \frac{1}{2}mv^2$ foot-poundals

where, m = mass of the car in pounds

v = speed of the car in ft/sec.

Putting the given values of m and v, we get work done

work done $= \frac{1}{2} \times 2 \times 2240 \times 132^2$ foot-poundals

[1 ton = 2240 lbs. 90 mph = 132 ft/sec]

$$= \frac{1}{2} \times \frac{2 \times 2240 \times 132^2}{32} \text{ ft pounds}$$

[1 ft pound = 32 ft poundals]

$$\therefore \text{Heat produced} = Q = \frac{2 \times 2240 \times 132^2}{2 \times 32 \times 1400} \text{lb.}^\circ\text{C.}$$

[1lb. °C in the heat required to raise the temperature of 1 lb by water through 1°C]

But Q = m.s.t.

$$t = \frac{Q}{m - s.} = \frac{2 \times 2240 \times 132^2}{1400 \times 2 \times 32 \times 64 \times 0.14} = 97.3°C$$

[m = 4 × 16 = 64 lbs (Total mass of brakes)]

Problem 2 : A river flowing horizontally, reaches a vertical fall of 100 meters. If the temperature of the water at the top is 20°C. What will be the water temperature

(a) half-way down the fall?

(b) in the pool at the bottom of the fall?

Solution :

(a) Let us consider 1 g of water at the top of the fall,

Then potential energy lost by it, when it is half-way down the fall, completely changes into kinetic energy i.e. mechanical energy of one form (P.E.) changes into mechanical energy of another form (K.E.). Thus no loss of mechanical energy has occurred so no heat is obtaineld and so the temperature of water will be same as at top i.e. 20°C.

(b) When 1 g of water reaches the bottom on the fall the potential energy because zero (∴ h = 0) and K.E. also becomes zero (its speed becomes practically zero). Thus mechanical energy has been lost and an equivalent amount of heat is produced.

Mechanical energy lost = Potential energy at the top of the fall

$$= mgh$$

$$= 1 \times 980 \times (100 \times 100) \text{ ergs.}$$

$$= \frac{9800000}{10^7} \text{ Joules}$$

$$\therefore \quad \text{Heat produced} = \frac{9800000}{10^7 \times 4.2} \text{ cal} = 0.23 \text{ cal.}$$

∴ temp. rise in 1 g = 0.23 °C

∴ Temp. of water in pool = 20 + 0.23 = 20.23°C

Problem 3 : In the absence of bearing friction a winding engine could raise a cage weighing 1000 kg. at 10 m/s, but this is reduced by friction of 9 m/s. How much oil, initially at 20°C, is required per second to keep the temperature of the bearings down to 70°C?

[sp. heat of oil = 0.5; g = 981 cm/s^2; J =4.2×10^7 ergs/calorie]

Solution : To find the mass of oil per second let us calculate the work done per second

In 1 sec.

work done without friction = force × distance

$$=10^6 \times 981 \times 1000 \text{ ergs.}$$

Work done in one sec with friction present

$$=10^6 \times 981 \times 900 \text{ ergs.}$$

∴ Work done against friction is IS = $10^6 \times 981$ (1000–900) ergs.

∴ Heat produced per second $= \dfrac{10^6 \times 981 \times 100}{4.2 \times 10^7}$ calories.

If the mass of oil required = mg/s

Then the heat given to the oil per sec = m.s.t

$$= m \times 0.5 \times (70-20)$$

$$\therefore\ m \times 0.5 \times (70-20) = \frac{10^6 \times 981 \times 100}{4.2 \times 10^7}$$

or $$m = \frac{10^6 \times 981 \times 100}{4.2 \times 10^7 \times 0.5 \times 50}\ \text{g}$$

$$= 93.5 \text{ g.}$$

Problem 4 : How much heat is used in performing on external work when 1 g of water of 100°C is boiled away to form steam at 100°C.

[density of water = 1g/cc; density of steam = 0.606 g/l; atmospheric pressure = 76 cm of mercury)

Solution : Volume of 1 g water at 100°C = 1 cc.

Volume of 1 g of steam at 100°C $= \dfrac{1000}{0.606}$ cc.

$\therefore$ Expansion in volume $= \left(\dfrac{1000}{0.606} - 1\right)$ cc.

This expansion takes place against the constant pressure of 1 atmosphere (i.e. 76×13.6×980 dynes/sq cm)

$\therefore$ Work done =- Pressure × change in volume

$= 76 \times 13.6 \times 980 \times 1649$ ergs

$$\therefore \text{ Heat required} = \frac{76 \times 13.6 \times 980 \times 1649}{4.2 \times 10^7} \text{ cal}$$

$= 39.8$ cal

Problem 5 : The enthalpies of a certain gas before and after suffering Joule-Thomson expansion through a throttle valve are 77.2 cal and 106.8 cals respectively. Deduce the fraction of the gas liquified if the enthalpy of emerging liquid be 55.4 calories.

Solution : The fraction of a gas liquified in a regeneration system is given by

$$X = \frac{H_f - H_e}{H_f - H_l}$$

Given $H_e = 77.2$ cal; $H_f = 106.8$ cal, $H_L = 55.4$ cal

$$\text{Thus} \quad X = \frac{106.8 - 77.2}{106.8 - 55.4} = \frac{29.6}{51.5}$$

or x = 0.57

10

Entropy

Problem 1 : Calculate the change in entropy when 10g of ice is at 0°C converted into water of the same temperature, (latent heat of ice = 80 cal/gm)

[Delhi 1975]

Solution : Using the relation, $ds = \frac{dQ}{T}$

we get, $dQ = 10 \times 80 = 800\text{cal}$

$T = 0°C = 273 \text{ K}$

$$\therefore ds = \frac{800}{273} = 2.93 \text{ Cal / K}$$

Problem 2 : Calculate the change of entropy when 1g atom solid mercury of its melting point is raised to a temperature of 40°C. [Given for mercurry –M.P. = –39°C, Latent heat of fusion = 3.00 cal/g, mean sp. heat = 0.0335 cal/g – K and one gm atom of mercurry = 200 gm].

[Agra 1971]

Solution : To calculate the change in entropy this process may be considered to take place in two steps

(i) Mercury changes from solid to liquid state at its M.P. i.e. –39°c.

(ii) Temperature of molten mercury increases from –39°C to 40°C.

Change in entropy when solid mercury changes to liquid state at its M.P.

$$\Delta S_1 = \frac{\Delta Q}{T} = \frac{mL}{T}$$

$$= \frac{200 \times 3.0}{234} \qquad [-39°C = -39 + 273 = 234K]$$

$$= 2.564 \text{ cal / K}$$

change in entropy when molten mercurry rises in temp. from –39°C to 40°C

$$\Delta S_2 = \int_{T_1}^{T_2} \frac{dQ}{T} = mc \int_{T_1}^{T_2} \frac{dT}{T}$$

$$= mc \ln \frac{T_2}{T_1} = 2.303\, mc \log \frac{T_2}{T_1}$$

$$= 2.303 \times 200 \times 0.0335 (\log 313 - \log 234)$$

$$= 2.303 \times 200 \times 0.0335 (2.4955 - 2.3692)$$

$$= 2.303 \times 200 \times 0.0335 \times 0.1263$$

$$= 1.948 \text{ cal/K.}$$

Hence Total increase in entropy

$$\Delta S = \Delta S_1 + \Delta S_2$$

$$= (2.564 + 1.948) \text{ cal/K}$$

$$= 4.512 \text{ cal/K.}$$

Problem 3 : The melting point of tin is 232°C, its latent heat of fusion 14 cal/gm and the specific heat of solid and molten tin 0.055 and 0.064 cal/gm respetively. Calculate the change in entropy when 1g of tin is heated from 150°C to 314°C.

[Agra 1976]

Solution : The process is completed in following three steps:

(i) Solid tin is heated from 150°C to 232°C

(ii) Solid tin changes to liquid tin at its M.P. (232°C)

(iii) Temperature of molten tin increases from 232°C to 314°C.

Change in entropy in each step can now be calculated.

Change in entropy when solid tin is heated

$$\Delta S_1 = \int_{423}^{505} \frac{dQ}{T}$$

$$\Delta S_1 = \int_{423}^{505} \frac{mc_v dt}{T}$$

$$= mc_v = \int_{423}^{505} \frac{dT}{T} \; 1 \times 0.055 \times \int_{423}^{505} \frac{dT}{T}$$

$= 2.303 \times 1 \times 0.055 (\log 505 - \log 423)$
$= 2.303 \times 1 \times 0.055 (2.7033 - 2.6263)$
$= 2.303 \times 1 \times 0.055 \times 0.0770$
$= 0.0097$ cal/K.

Change in entropy when molten tin is obtained from solid tin at 505K.

$$\Delta S_2 = \frac{dQ}{T} = \frac{m \times L}{T}$$

$$= \frac{1 \times 14}{505} = 0.028 \; cal / K$$

Change in entropy when temperature of molten tin increases from 505K to 587K.

$$\Delta S_3 = \int_{505}^{587} mc_v \frac{dT}{T}$$

$= 1 \times 0.064 \times 2.303 (\log 587 - \log 505)$
$= 1 \times 0.064 \times 2.303 (2.7686 - 2.7033)$
$= 1 \times 0.064 \times 2.303 \times 0.0653$
$= 0.0096$ cal/K.

∴ Total change in entropy

$\Delta S = \Delta S_1 + \Delta S_2 + \Delta S_3$
$= 0.0097 + 0.028 + 0.0096$
$= 0.047$ cal/K.

Problem 4 : Calculate the increase in entropy when 40g of water at 20°C is converted into vapour at 100°C. [Given: sp. heat of water = 4.2 J gm^{-1} $degree^{-1}$, Latent heat of water = 2.27×10^3 J gm^{-1}.]

Solution : The entropy increases in following steps.

(i) When temp. of 40g of water increases from 40°C to 100°C (ΔS_1)

(ii) When 40g of water at 100°C gets converted into vapours at constant temperature (100°C), ΔS_2

Now,

$$\Delta S_1 = \int_{T_1}^{T_2} \frac{dQ}{T} = mc_v \int_{T_1}^{T_2} \frac{dT}{T}$$

$$= 40 \times 4.2 \times 2.303 (\log 373 - \log 293)$$

$$= 40 \times 4.2 \times 2.303 (2.5717 - 2.4669)$$

$$= 19.44 \text{ J/K.}$$

Again

$$\Delta S_2 = \frac{dQ}{T} = \frac{mC}{T}$$

or $$\Delta S_2 = \frac{40 \times 2.27 \times 10^3}{373}$$

$$= 243.42 \text{ J/K.}$$

Hence total Increase in entropy,

$$\Delta S = \Delta S_1 + \Delta S_2$$

$$= (19.44 + 243.42) \text{ J/K}$$

$$= 262.86 \text{ J/K.}$$

Problem 5 : 10g of water at a temperature of 20°C is converted into ice at –10°C at constant atmospheric pressure. Assuming specific heat of ice at constant pressure to be 0.5 cal per gram per °C, find the total entropy change of the system. Latent heat of ice in 80 cal/gm

[Agra 1973]

Solution : The various steps in which change in entropy occurs are:

(i) Temperature of water changes from 20°C to 0°C (entropy decreases)

(ii) Water changes into ice at 0°C (entropy decreases)

(iii) ice is cooled from 0°C to –10°C (entropy decreases)

Now,

Decrease in entropy in step (i)

$$\Delta S_1 = \int_{T_1}^{T_2} \frac{dQ}{T} = 2.303 \times mc \times (\log 293 - \log 273)$$

$= 2.303 \times 10 \times 1 \times (2.4669 - 2.4362)$

$= 2.303 \times 10 \times 1 \times (0.00307)$

$= 0.7069$ cal/K.

Decrease in entropy in step (ii)

$$\Delta S_2 = \frac{dQ}{T} = \frac{mc}{T}$$

$$= \frac{10 \times 80}{273} = 2.9304 \text{ cal / K}$$

Decrease in entropy in step (iii)

$$\Delta S_3 = \int_{263}^{273} \frac{dQ}{T} = 2.303 \times mc \times (\log 273 - \log 263)$$

$= 2.303 \times 14 \times 0.5 \times (2.4363 - 2.4200)$

$= 0.1865$ cal/K.

∴ Total decrease in entropy (ΔS)

$\Delta S = \Delta S_1 + \Delta S_2 + \Delta S_3$

$= 0.7069 + 3.9304 + 0.1865$

$= 3.8238$ cal/K.

Problem 6 : Calculate in increase in entropy when 1kg of ice at –10°C is converted into steam at 100°C. [Sp. heat of ice = 0.5 k cal/ kg/°C, latent heat of ice = 3.4×10^5 J/kg, Latent heat of steam = 22.68×10^5 J/kg; J = 4.2 Joule/calorie)

[Agra 1985]

Solution : The various steps in this change are:

(i) Increase in entropy when temperature of ice increases from –10°C to 0°C (ΔS_1)

(ii) Increase in entropy when ice melts at 0°C to give water at 0°C (ΔS_2)

(iii) Increase in entropy when temperature of water rises from 0°C to 100°C (ΔS_3)

(iv) Increase in entropy when water at 100°C changes to steam (ΔS_4)

The changes in entropy for various steps can be calculated as follows:

Increase of entropy in step (i)

$$\Delta S_1 = \int_{263}^{273} \frac{dQ}{T} = mc_v \int_{263}^{273} \frac{dT}{T}$$

$$= 1 \times 0.5 \times 2.303 (\log 273 - \log 263)$$

$$= 0.5 \times 2.303 \times 0.0162$$

$$= 0.01865 \text{ kcal/K}$$

$$= 18.65 \text{ cal/K.}$$

Increase in entropy in step (ii)

$$\Delta S_2 = \frac{dQ}{T} = \frac{mL}{T} = \frac{1 \times 3.4 \times 10^5}{273} \text{ J / kg}$$

$$= 1.245 \times 10^3 \text{ J/K}$$

$$= \frac{1.245 \times 10^3}{4.2} \text{ cal / kg} = 296.5 \text{cal / K.}$$

Increase in entropy in step (iii)

$$\Delta S_3 = \int_{273}^{373} \frac{dQ}{T} = mc \times 2.303 \left(\log 373 - \log 273\right)$$

$$= 1 \times 1 \times 2.303 \times 0.135$$

$$= 0.3109 \text{ k cal/K}$$

$$= 310.9 \text{ cal/K.}$$

Increases in entropy in step (iv)

$$\Delta S_4 = \frac{dQ}{T} = \frac{mL}{T} = \frac{22.68 \times 10^5}{373} \text{ J / K}$$

$$= 6.080 \times 10^3 \text{ J/K}$$

$$= 1447.72 \text{ Cal/K} \qquad [\quad 1 \text{ cal} = 4.2\text{J})$$

Thus total Increase in entropy, ΔS

$$\Delta S = \Delta S_1 + \Delta S_2 + \Delta S_3 + \Delta S_4$$
$$= (18.65 + 296.5 + 310.9 + 1447.7) \text{ cal/K}$$
$$= 2073.75 \text{ cal/K}$$

Problem 7 : 10kg of water at 0°C is mixed with an equal man of water at 80°C. Calculate the change in entropy. [sp. heat of water = 1k cal/kg –K]

Solution : Let the final temperature of mixture be t°C, then

$$10 \times 1 \times (t - 0) = 10 \times (80 - t)$$

[Heat gained = Heat lost]

or $10t = 800 - 10t$

or $20t = 800$

or $t = 40$

i.e., the final temperature of the mixture is 40°C, or 313K.

To calculate the change in entropy we have to consider the various steps involved in the process.

(i) **Increase** in entropy when 10kg of water rises from 273K to 313K (ΔS_1)

(ii) **Decrease** in entropy when the temperature of 10kg of water falls from 353K to 313K, (ΔS_2)

Increase in step (i)

$$\Delta S_1 = \int_{273}^{313} \frac{dQ}{T} = mc \int_{273}^{313} \frac{dQ}{T}$$

$$= 10 \times 1 \times 2.303 (\log 313 - \log 273)$$
$$= 10 \times 1 \times 2.303 (2.4955 - 2.4362)$$
$$= 10 \times 1 \times 2.303 \times 0.0593$$
$$= 1365.4 \text{ cal/K.}$$

Decrease in Entropy in Step (ii)

$$\Delta S_1 = \int_{313}^{353} \frac{dQ}{T} = mc \int_{313}^{353} \frac{dT}{T}$$

$$= 10 \times 1 \times 2.303 \text{ (log } 353 - \log\ 313)$$
$$= 10 \times 1 \times 2.303\ (2.5478 - 2.4955)$$
$$= 10 \times 1 \times 2.303 \times 0.0523$$
$$= 1204.2 \text{ Cal / K}$$

Thus the net change in entropy, ΔS is

$$\Delta S = \Delta S_1 - \Delta S_2$$
$$= 1365.4 - 1204.2$$
$$= 161.2 \text{ cal/K.}$$

Problem 8 : A mass m of a liquid at temperature T_1 is mixed with an equal mass of the same liquid at temperature T_2. The system is personally insulated. Show that the entropy change of the universe is. $2m\ c_p \ln \dfrac{T_1 + T_2}{2\sqrt{T_1 T_2}}$.

Solution : When a mass 'm' g of a at temperature T_1 is mixed with an equal mass of the same liquid at temperature T_2, the temperature of the mixture will be $\dfrac{T_1 + T_2}{2}$.

$\therefore$ change in entropy, ΔS_1, is given by

$$\Delta S_1 = \int_{T_1}^{\frac{T_1+T_2}{2}} \frac{dQ}{T} = \int_{T_1}^{\frac{T_1+T_2}{2}} m\ c_p \frac{dT}{T}$$

$$= m.c_p \ln \frac{T_1 + T_2}{2T_1}$$

Change in entropy also occurs when the temp. of m gm of liquid changes from T_2 to $\dfrac{T_1 + T_2}{2}$, then this change in entropy, ΔS_2, is given by

$$\Delta S_2 = \int_{T_2}^{\frac{T_1+T_2}{2}} m.c_p \frac{dT}{T} = m.c_p \ln \frac{T_1 + T_2}{2T_2}$$

Hence entropy change of universe is given by

$$\Delta S = \Delta S_1 + \Delta S_2$$

$$= m.c_p \ln \frac{T_1 + T_2}{2T_1} + m.c_p \ln \frac{T_1 + T_2}{2T_2}$$

$$= m.c_p \ln \left[\frac{T_1 + T_2}{2T_1} \times \frac{T_1 + T_2}{2T_2}\right]$$

$$= m.c_p \ln \left[\frac{(T_1 + T_2)}{2\sqrt{T_1 T_2}}\right]^2$$

$$= 2 \ . \ m.c_p \ln \frac{T_1 + T_2}{2\sqrt{T_1 T_2}}$$

This value will be **positive.**

Problem 9 : 10g of steam at 100°C is blown into 90g of water at 0°C contained in a calorimeter of water equivalent 10g. The whole of the steam is condensed. Calculate the increase in entropy of the system.

[Delhi 1973]

Solution : Let the final temperature of the mixture be t°C i.e. (273 + t) K.

Then

Heat lost by steam = 10 × 540 + 10 [373 – (273 + t)]

Heat gained by water

and colorimeter = (90 + 10) (273 + t) – 273)

Now Heat gained = Heat lost

$\therefore 10 \times 540 + 10\,(100 - t) = 100 \times t$

$5400 + 1000 - 10t = 100\ t$

or $6400 = 110\ t$

or $T = \dfrac{6400}{110}$

$= 58.19°C$ or $331.19K$.

The entropy change occurs in following steps.

(i) Increase in entropy when temperature of water rises from 273 to 331.19K. (ΔS_3)

(ii) Decrease in entropy when 10g of steam condenses to water at 373K. (ΔS_2)

(iii) Decrease in entropy when 10g of water cools from 373K to 331.19K. (ΔS_3)

Increase in entropy in step (i)

$$\Delta S_1 = \frac{dQ}{T} = mC \int_{273}^{331.19} \frac{dT}{T}$$

$$= 100 \times 1 \times 2.303 (\log 331.19 - \log 273)$$

$$= 19.324 \text{ cal/K.}$$

Change in entropy in step (ii)

$$\Delta S_2 = \frac{\Delta Q}{T} = -\frac{10 \times 540}{373} = -14.477 \text{ cal / K}$$

Change in entropy in step (iii)

$$\Delta S_3 = mC \int_{373}^{331.19} \frac{dT}{T}$$

$$= 10 \times 1 \times 2.303 (\log 331.19 - \log 373)$$

$$= 10 \times 1 \times 2.303 (2.5202 - 2.5717)$$

$$= -10 \times 2.303 \times 0.0515$$

$$= -1.188 \text{ cal/K.}$$

Total change in entropy, ΔS is

$$\Delta S = \Delta S_1 + \Delta S_2 + \Delta S_3$$

$$= 19.324 - 14.477 - 1.188$$

$$= +3.659 \text{ cal/K.}$$

Problem 10 : Calculate the heat required to convert 1g of water into steam under standard atmospheric pressure if the entropy of water at the boiling point is 0.31 and for 1000 steam at the steam temperature is 1.74

[Agra 1980]

Solution : We know, $dS = \frac{dQ}{T}$

or $dQ = T.ds$

$\therefore \int dQ = T\int dS$

or $Q = T(S_2 - S_1)$

$= 373\ (1.74 - 0.31)$

$= 533.4$ cal/g.

Problem 11 : One gram molecule of a perfect gas expands isothermally to four times its initial volume. Calculate the change in its entropy in term of the gas constant, R.

[Rohilkhand 1982]

Solution : Work of expansion is given by

$$W = \int_{v_1}^{v_2} P dv$$

But for a perfect gas PV = RT

$$\text{or} \quad \overline{V} = \frac{RT}{P}$$

$$\therefore \overline{W} = RT \int_{v_1}^{v_2} \frac{dv}{v}$$

$$= 2.303\ RT \log \frac{V_2}{V_1}$$

Given: $v_1 = v$; $v_2 = 4v$

$$\therefore \frac{V_2}{V_1} = \frac{4V}{V} = 4$$

Hence W = 2.303 RT log 4

If the heat used for doing his work is d then change in entropy,

$$ds = \frac{dQ}{T} = \frac{w}{T} = -\frac{2.303 \times RT \times 0.62}{T}$$

$$= 1.386R\ J/K\cdot$$

Problem 12 : 5g of nitrogen doubles in volume at constant temperature. Find out the change in entropy.

Solution : $\Delta S = S_2 - S_1 = 2.303\ C_v \log \frac{T_2}{T_1} + (C_p - C_v)$

$$\times\ 2.303 \log \frac{V_2}{V_1}.$$

In this case there is no change in temperature and the mass of the gas is 5g

$$\therefore S_2 - S_1 = 5\,(C_p - C_v) \times 2.303 \log \frac{V_2}{V_1}$$

$$\left[\because \log \frac{T_2}{T_1} = 0\right]\left[\text{as } T_2 = T_1\right]$$

$$= 5 \times V \times 2.303 \log \frac{V_2}{V_1}$$

[Where γ = gas constant for 1g]

For nitrogen $\gamma = \frac{R}{28}$

$$= \frac{76 \times 13.6 \times 981 \times 22400}{28 \times 273}\left[R = \frac{PV}{T}\right]$$

$$= \frac{8.3 \times 10^7}{28}\ \text{ergs/K}\ \frac{8.3}{28}\ \text{J/K}$$

$$\therefore \text{Change in entropy } S_2 - S_1 = 5 \times \frac{8.3}{28} \times 2.303 \log 2$$

$= 1.035$ J/K

Problem 13 : For silver, molar specific heat at constant pressure in the range 50 to 100K is given by $C_p = 0.076T - 0.00026T^2 - 0.15$ cal/K. where T is the Kelvin temperature. If 2 moles of silver is heated from 50 to 100K. Calculate (i) heat required (ii) change entropy.

[Agra 1977]

Solution : (i) Heat Required

We know $dQ = C_p.dt$ (for 1 mole)

$\therefore dQ = 2 \times C_p\ dt$ (for 2 moles)

$= 2\,(0.076T - 0.00026T^2 - 0.15)\ dT$

The temperature increases from 50 k to 100K

$$\therefore\ Q = \int_{50}^{100} dQ = \int_{50}^{100} 2\left(0.076T - 0.00026T^2 - 0.15\right)DT$$

$$= 2\left[\frac{0.076T^2}{2} - \frac{0.00026T^3}{3} - 0.15T\right]_{50}^{100}$$

$$= 2\left[\frac{0.076\left(100^2 - 50^2\right)}{2} - \frac{0.00026\left(100^3 - 50^3\right)}{3} - 0.15\left(100^3 - 50^3\right)\right]$$

= 403.9 calories.

(ii) change in entropy when silver is heated from 50 k to 100K.

$$\Delta S = S_2 - S_1 = \int_{50}^{100} \frac{dQ}{T}$$

$$= \int_{50}^{100} 2\left(\frac{0.076T - 0.00026T^2 - 0.15}{T}\right)dt$$

$$= \int_{50}^{100} 2\left(0.076T - 0.00026T - \frac{0.15}{T}\right)dt$$

$$= 2\left[0.076T - \frac{0.00026T^2}{T} - 0.15\ln T\right]_{50}^{100}$$

$$= 2\left[0.076\ (100 - 50) - 0.00026\left(\frac{100^2 - 50^2}{2}\right) - 0.15\ln\frac{100}{50}\right]$$

= 2 × 0.076 × 50 – 6.00026 × 150 × 50 – 2 × 0.15 × 2.303 by2

= 3.58 cal/k.

11

Thermo Dynamic Relationships

Problem 1 : Calculate the latent heat of ice given that change of pressure by atmosphere changes the melting point of ice by 0.0074°C and when one gm of ice melts its volume changes by 0.0907 cc.

[Rohilkhand 1984s]

Solution : Using the Claussius–Clapeyron equation, we have

$$\frac{dP}{dT} = \frac{L}{T(V_2 - V_1)}$$

$$L = \frac{dP, T(V_2 - V_1)}{dT}$$

Given:

dp = 1 atmosphere; 1.1013×10^6 dynes/cm^2

$T = 0°c = 273k$; $V_2 - V_1 = 0.0907$ cc

at = 0.0074°C = 0.0074°k

Putting these values in above equation we get

$$L = \frac{1.013 \times 10^6 \times 273 \times 0.0907}{0.0074}$$

$$= 3.39 \times 10^9 \text{ ergs / gm} = \frac{3.39 \times 10^9}{4.18 \times 10^7} \text{ cal / gm.}$$

$$= 81.1 \text{ cal/gm.}$$

Problem 2 : Calculate the change in boiling point of water when the pressure is increased from 1.0 to 1.2 atmosphere. [Given : Specific volumes of steam = 1677cc/gm; Latent heat of steam = 540cal/gm; Boiling point of water at atmosphere pressure = 100°C; and Pressure of 1 atmosphere = 1.0×10^6 dyves/cm^2]

[Agra 1970]

Solution : Using Clapeyron–Claussius equation, we have

$$dT = \frac{T(V_2 - V_1)}{2} dP$$

Given : dP = 1.2 – 1 = 0.2 atmosphere = $0.2 \times 1 \times 10^6$ dynes/cm^2

$T = 100°C = 373k$

$V_2 - V_1 = (1677 - 1) = 1676$ cc/gm

$L = 540$ cal/g $= 540 \times 4.2 \times 10^7$ ergs/g

∴ Change in boiling point of water, dT, can be obtaining by subtituting the given values,

$$\therefore dT = \frac{373 \times 1676 \times 0.2 \times 10^6}{540 \times 4.2 \times 10^7}$$

$$= 5.5 \text{ k or } 5.5°C$$

Problem 3 : Calculate the change in melting point of wax for a pressure of 50 atmospheres from the following data:

Melting point = 64°C; volume at 0°C = 1cc.

Volume of solid at melting point = 1.161 cc.

Volume of liquid at melting point = 1.166 cc.

Density of solid at 64°C = 0.96g/cc.

Latent heat = 97cal/gm.

[Agra 1961]

Solution : The melting point of solid at atmospheric pressure is 64°C but we are required to find change in melting point for a pressure of 50 atmospheres. i.e. for a pressure change of 49 (50 – 1 = 49) atmospheres.

∴ $dP = 49 \times 76 \times 13.6 \times 981$ dynes/cm^2

$T = 64°C = 337k$

$L = 97 \text{ cal} = 97 \times 4.2 \times 10^7$ ergs.

Mass of solid at melting point = volume x density

$= 1.161 \times 0.96$ gm.

$\therefore$ Specific volume of the solid at m.p, V_1 is

$$V_1 = \frac{\text{volume}}{\text{mass}} = \frac{1.161}{1.161 \times 0.96} = 1.0417 \text{ cc./g}$$

and specific volume of liquid at m.p. V_2 in

$$V_2 = \frac{\text{volume}}{\text{mass}} = \frac{1.166}{1.166 \times 0.96} = 1.0461 \text{ cc./g}$$

$$\therefore V_2 - V_1 = 1.0461 - 1.0417 = 0.0044 \text{ cc/g}$$

Now using Clapeyron–Claussius equation, we get

$$\frac{dP}{dT} = \frac{L}{T(V_2 - V_1)}$$

or
$$dT = \frac{dP \times T(V_2 - V_1)}{L}$$

Substituting the volues in above equation,

$$dT = \frac{49 \times 76 \times 13.6 \times 981 \times 337 \times 0.0044}{97 \times 4.2 \times 10^7}$$

$= 0.018$k. or 0.018°C.

Problem 4 : Calculate the depression in the melting point of ice by one atmospheic increase of pressure, given that the latent heat of fusion of ice $L = 3.4 \times 10^5$ Joules/kg. Specific volume of 1kg of ice and water at 0°C are $1.091 \times 10^{-3}\text{m}^2$ and $1.00 \times 10^{-3}\text{m}^2$ respectively. (1 atmospheric pressure = 10^5 N/m^2)

[Raj. 1986]

Solution : Given :

$dP = 1$ atmosphere $= 10^5$ N/m^2

$T = 0°C = 273$k

$L = 34 \times 10^5$J/kg

Specific volume of 1kg ice, $v_1 = 1.091 \times 10^{-3}\text{m}^3$

Specific volume of 1kg water, $v_2 = 1.000 \times 10^{-3} m^3$

Thus, $v_2 - v_1 = (1.000 \times 10^{-3}) - (1.091 \times 10^{-3}) m^3$

$= -0.091 \times 10^{-3}$

Using clapeyron equation,

$$= \frac{dP}{dT} = \frac{L}{T(v_2 - v_1)}$$

We get $$dT = \frac{dP \times T(v_2 - v_1)}{L}$$

on $$dT = \frac{10^5 \times 273(-0.091 \times 10^{-3})}{3.4 \times 10^5}$$

$= -0.0073k$ or $-0.0073°C$

Problem 5 : Calculate the depression of melting point of ice (L = 80 cal) per atmosphere increase of pressure, if the ratio of densities of ice and water at 0°C is $\frac{10}{11}$.

[Allahabad 1989]

Solution : Using Claussius–Clapeyron equation,

$$dT = \frac{dP \times T(v_2 - v_1)}{L}$$

Given: T = 273k; L = 80 cal = $80 \times 4.2 \times 10^7$ ergs

dP = 1 atmosphere = 1.03×10^6 dynes/cm^2

$$\frac{v_2}{v_1} = \frac{\text{specific volume of water}}{\text{specific volume of ice}}$$

$$= \frac{\text{Density of ice}}{\text{Density of water}}$$

$$= \frac{10}{11}$$

But $v_2 = 1.0$ cm^3/g

$$\therefore v_1 = \frac{11}{10} \times 1.0 = 1.1 \text{ cm}^3/\text{g}$$

$$\therefore v_2 - v_1 = 1.0 - 1.1 = -0.1 \text{ cm}^3/\text{g}$$

$$\text{Hence } dT = \frac{273 \times (-0.1) \times 1.013 \times 10^6}{80 \times 4.2 \times 10^7}$$

$$= -0.0082°C \text{ or } -0.0082 \text{ k}$$

So decrease in m.p., of ice is 0.0082°C per atmosphere increase of pressure.

Problem 6 : Calculate the pressure required to make the ice freeze at –1°C,. change of specific volume when 1g of water freezes into ice = 0.091cc J = 4.2×10^7 ergs/cal; 1 atmosphere = 10^6 dynes/cm^2 and the Latent heat of ice = 80 cal/g.

[Agra 1985]

Solution : Given: $dT = -1°C$; $v_2 - v_1 = -0.091$cc.

$L = 80 \text{ cal} = 80 \times 4.2 \times 10^7$ ergs

$T = 0°C = 273$ k.

Using Calapeyron's equation

$$\frac{dP}{dT} = \frac{L}{T(v_2 - v_1)} \quad \text{or} \quad dP = \frac{L\, dT}{T(v_2 - v_1)}$$

we get,

$$dP = \frac{80 \times 4.2 \times 10^7 \times (-1)}{273 \times (-0.091)} \text{ dynes/cm}^2$$

$$= \frac{80 \times 4.2 \times 10^7}{273 \times 0.091 \times 10^6} \text{ atmospheres}$$

$$= 135.2 \text{ atmospheres}$$

It means that to lower the melting point of ice by 1°C we must raise the pressure by 135.2 atmospheres. i.e., we could require a pressure of 136.2 (1 + 135.2 = 136.2) atmosphere if we wish to make freeze ice at –1°C (because the ice freezes at 0°C at 1 atmospheric pressure)

Problem 7 : Claculate the change in vapour pressure of water as the boiling temperature changes from 100°C to 103°C. Given the latent heat of steam = 540 cal/gm and specific volume of steam = 1670cc/gm.

[Agra 1976]

Solution : Given: $L = 540 \text{ cal} = 540 \times 4.2 \times 10^7 \text{ergs}$

$T = 100°C = 373 \text{ k}$

$dT = (103 - 100) = 3°C$

$(v_2 - v_1) = (1670 - 1) \text{ cc} = 1669 \text{ cc.}$

Using Clapeyron equation

$$\frac{dP}{dT} = \frac{L}{T(v_2 - v_1)}$$

or
$$dP = \frac{L\, dT}{T(v_2 - v_1)}$$

Substituting the given values, we get

$$dP = \frac{540 \times 4.2 \times 10^7 \times 3}{373 \times 1669} \text{ dynes/cm}^2$$

$$= \frac{540 \times 4.2 \times 10^7 \times 3}{373 \times 1669 \times 76 \times 13.6 \times 981} \text{ atmospheres}$$

$$= 0.17091 \text{ atmospheres}$$

Problem 8 : Water boils at 100.8°C and 99.8°C when the atmospheric pressures are respectively 79.371 and 76.650 cm of mercurry. Calculate the volume of 1g steam at 100°C. The latent heat being 540 cal/g.

Solution : Given: $dT = (100.8 - 99.8) = 1°C$

$L = 540 \text{ cal} = 540 \times 4.2 \times 10^7 \text{ ergs}$

$T = 100°C = 373k$

$dP = 79.371 - 76.650 = 2.721 \text{ cm of Hg}$

$= 2.721 \times 13.6 \times 981 \text{dynes/cm}^2$

Using Clapeyron equation

$$\frac{dP}{dT} = \frac{L}{T(v_2 - v_1)}$$

we have
$$(v_2 - v_1) = \frac{L\, dT}{T\, dP}$$

Substituting the given values, we get

$$(v_2 - v_1) = \frac{540 \times 4.2 \times 10^7 \times 1}{373 \times 2.721 \times 13.6 \times 981}$$

$$= 1669 \text{ cc}$$

But $v_1 = 1$cc

$\therefore v_2 = 1669 + 1 = 1670$ cc.

Hence volume of 1g of steam at 100°C = 1670 cc.

Problem 9 : Calculate the specific heat of saturated steam at 100°C from the following data :

L at 90°C = 545.25 cal

L at 100°C = 539.30 cal

L at 110°C = 533.17 cal

Specific heat of water at 100°C = 1.013 cal/gm

[Delhi 1975]

Solution : $C_1 = 1.013$ cal/gm; $C_2 = ?$

$$\text{Now, } \frac{dL}{dT} = \frac{\text{L at } 110°\text{C} - \text{L at } 90°\text{C}}{20} = \frac{533.17 - 545.25}{20}$$

$$= -\frac{12.08}{20} = -0.604 \text{ cal / gm}$$

$$\text{Thus, } c_2 - c_1 = \frac{dL}{dT} - \frac{L}{T}$$

$$\text{or } c_2 = c_1 + \frac{dl}{dt} - \frac{L}{T}$$

$$= 1.013 - 0.604 = \frac{539.30}{373}$$

$$= -1.037 \text{ cal/gm.}$$

Problem 10 : If L = 800 – 0.705T, show that the specific heat of steam is negative.

[Kanpur 1979]

Solution : Given: L = 800 – 0.705T ...(i)

$$\therefore \frac{dL}{dT} = -0.705 \text{ [differentiating (i) w.r.t.T]}$$

B.P. of water 100°C under 1 atmosphere pressure so Latent heat L = 800 – 0.705 × 373 at B.P. (373k)

Sp. heat of water (liquid), $c_1 = 1$

Using the equation

$$c_2 - c_1 = \frac{dL}{dT} - \frac{L}{T}$$

we get $$c_2 = c_1 + \frac{dL}{dT} - \frac{L}{T}$$

$$= 1 + (-0.705) - \frac{537}{373}$$

$$= 1 + (-0.705) - 1.44$$

$$= 1 - 0.705 - 1.44 = -1.145$$

It shows that the specific heat of water at 100°C is negative

Problem 11 : The Latent heat of water diminishes by 0.695 cal/gm for each degree contigrade rise in temperature at a temperature 100°C and latent heat of water vapour at 100°c is 540 cal/gm. Calculate the specific heat for saturated steam.

[Kanpur 1977]

Solution : We know $c_2 - c_1 = \frac{dL}{dT} - \frac{L}{T}$

Given : $\frac{dL}{dT} = -0.695$ cal / gm, °C; T = 100°C = 373k

L = 540 cal/gm; $c_1 = 1.0$ cal/gm [For liquid water]

$$c_2 = 1 + (-0.695) - \frac{540}{373}$$

$$= 1 - 0.695 - 1.448$$

$$= -1.143 \text{cal/gm°C}$$

12

Classical Statistics

Problem 1 : Eight similar coins are tossed for a large number of times, calculate.

(a) The probability of getting the neads of 5 coins uppermost

(b) The probability of most probable combination

(c) The probability of least probable combination

[Rohtak 1977]

Solution :

The probability of the combination $a^r b^s$ is given by,

$$P = \frac{n!}{r!\, s!} \times \frac{1}{2^n} \text{ where } s = n - r$$

(a) In this case n = 8, r = 5 and s = 3

$$\therefore \text{Probability} = \frac{8!}{5!\ 3!} \times \frac{1}{2^8} = \frac{7}{32}$$

(b) In the case of most probable combination

$$r = \frac{n}{2} = \frac{8}{2} = 4$$

Therefore the probability of most probable combination is given by

$$P_{max} = \frac{8!}{(4!)^2} \times \frac{1}{2^8}$$

$$= \frac{5 \times 6 \times 7 \times 8}{1 \times 2 \times 3 \times 4} \times \frac{1}{2^8}$$

$$= \frac{70}{256} = \frac{35}{128}$$

(c) The probability of least probable state is given by

$$P_{min} = \frac{1}{2^n} \times \frac{1}{2^8} = \frac{1}{256}$$

Problem 2 : Calcultate the probability that in tossing a coin 10 times we get

(i) all heads
(ii) 5 heads, 5 tails
(iii) 3 heads, 7 tails
(iv) 7 heads, 3 tails

[Agra 1977]

Solution : We know that probability, P, is given by

$$P(r,s) = \frac{n!}{r!s!} \times \frac{1}{2^n} \qquad \text{Where } s = n-r$$

(i) Here n = 10, r = 10, s = 0.

$$\therefore\ P(10,0) = \frac{10!}{10!\ 0!} \times \frac{1}{2^{10}} = \frac{1}{2^{10}} = \frac{1}{1024}$$

(ii) Here n = 10, r = 5, s = 5

$$\therefore\quad P(5,5) = \frac{10!}{5!\ 5!} \times \frac{1}{2^{10}}$$

$$= \frac{10 \times 9 \times 8 \times 7 \times 6}{1 \times 2 \times 3 \times 4 \times 5} \times \frac{1}{2^{10}} = \frac{255}{1034}$$

(iii) Here n = 10, r = 3, s = 7

$$\therefore\quad P(3,7) = \frac{10!}{3!\ 7!} \times \frac{1}{2^{10}}$$

$$= \frac{10 \times 9 \times 8}{1 \times 2 \times 3} \times \frac{1}{2^{10}} = \frac{120}{2^{10}} = \frac{120}{1024}$$

(iv) Here n = 10, r = 7, s = 3

$$\therefore\ P(7,3) = \frac{10!}{7!\ 3!} \times \frac{1}{2^{10}} = \frac{120}{1024}.$$

Problem 3 : If P(r) in the probability of getting r heads and (N – r) tails in tossing N coins, P(s) is the probability of getting s heads and M–s tails is tossing M coins. State the probability of getting r heads out of N tosses and s heads out of M tosses simultaneously in one pair of trials.

Solution :

P(r) = Probability of getting r heads and (N – r) tails in tossing N coins.

P(s) = Probability of getting s heads and (M – s) tails in tossing M coins.

Since the probability of any composite event is equal to the product of probabilities of individual and independent of component events, therefore the probability of getting r heads out of N tosses and s heads out of M tosses simultaneously in one pair of trails

$$= (P(r) \times P(s))$$

Problem 4 : Two similar discs A and B each having six equally likely faces are marked as 1, 2, 3, 4, 5, 6 are thrown simultaneously. Calculate the probability of getting the faces of both the dice up marked with same number.

Solution : The probability of A getting the face marked 1 up $= \frac{1}{6}$ Since the dice has equally likely six faces. Similarly the probability of B getting the face marked 1 up $= \frac{1}{6}$

Since the probability of a composite event is equal to the product of probabilities of individual and independent component events. Therefore the probability of getting the faces of both the dice marked with number 1 upwards $= \frac{1}{6} \times \frac{1}{6} = \frac{1}{36}$

Similarly the probability of getting the faces marked '2' upwards of both the dice simultaneously $= \frac{1}{36}$

and same is true for faces marked 3, 4, 5, 6.

Thus we have 6 favourable independent events in which the dice will leave the faces marked with the same numbers upwards. Hence probability of getting the faces of both the dice marked with the same number upwards is

$$\frac{1}{36}+\frac{1}{36}+\frac{1}{36}+\frac{1}{36}+\frac{1}{36}+\frac{1}{36}=\frac{1}{6}$$

Problem 5 : If twelve particles are distributed randomly between two boxes A and B with equal probability, then calculate.

(a) The probability of the distribution (8, 4).

(b) The probability of most probably distribution.

(c) The probability of least probable distribution.

Solution : We know

$$P(r,s)=\frac{n!}{r!\ s!}\times\frac{1}{2^n} \qquad \text{where } s = n-r$$

(a) In this case r = 8, s = 4, n = 12

$$\therefore\ P(8,4)=\frac{12!}{8!\ 4!}\times\frac{1}{2^{12}}=495\times 2^{-12}$$

(b) The most probable distribution $r=\frac{n}{2}=\frac{12}{2}=6$

$$\therefore\ n = 12,\ r = 6,\ s = 6$$

Hence $$P_{max}=\frac{12!}{6!\ 6!}\times\frac{1}{2^{12}}=924\times 2^{-12}$$

(iii) Least probable distribution

$$P_{min}=\frac{1!}{2^n}=\frac{1}{2^{12}}=2^{-12}$$

Problem 6 : Find the probability that in tossing a coin 12 times we get. (i) 4 heads, 8 tails (ii) 6 heads, 6 tails.

[Rohilkhand 1984]

Solution :

$$P(r,s) = \frac{n!}{r!\ s!} \times \frac{1}{2^n}$$

(i) Here n = 12, r = 4, s = 8

$$\therefore\ P(4,8) = \frac{12!}{4!\ 8!} \times \frac{1}{2^{12}} = 495 \times 2^{-12}$$

(ii) Here n = 12, r = 6, s = 6

$$\therefore\ P(6,6) = \frac{12!}{6!\ 6!} \times \frac{1}{2^{12}} = 924 \times 2^{-12}$$

Problem 7 : A system of n particles has only two allowed states A and B. The probability for state A is P and for state B is (1 – P). What is the probability for system to be in macrostate defined by distribution [r, (n – r)]

[Rohilkhand 1984]

Solution : The probability of state A is p and that for state B is (1 – p). Thus the weight factors 'a' and 'b' for states A and B are p and (1–p) respectively. Therefore the probability for the distribution of n particles in states A and B in macrostate (r, n – r) is given by

$$P[r_1(n-r)] = \frac{n!}{r!\ (n-r)!} p^r.(1-p)^{n-r}$$

Problem 8 : Assume that each face of a six faced dice is equally likely to land uppermost. Consider a game which involves the tossing of five such dice. Calculate the probability that number 4 appears uppermost.

(i) in exactly 1 dice

(ii) in exactly 2 dice

(iii) in exactly 5 dice

(iv) in done of the 5 dice

(v) in at least one dice

[Agra 1982]

Solution : We know probability

$$P(r) = \frac{n!}{r!\,(n-r)!} \cdot \left(\frac{1}{6}\right)^r \left(\frac{5}{6}\right)^{n-r}$$

When r out of n dice show number '4' uppermost.

(i) $n = 5, r = 1$

$$\therefore \; P(r) = \frac{5!}{1!(5-1)!}\left(\frac{1}{6}\right)^1 \left(\frac{5}{6}\right)^{5-1}$$

$$= \frac{5!}{4!} \cdot \left(\frac{1}{6}\right)\left(\frac{5}{6}\right)^4 = \left(\frac{5}{6}\right)^5 = 0.4$$

(ii) $n = 5, r = 2$

$$\therefore \quad P(2) = \frac{5!}{2!\,(5-2)!} \cdot \left(\frac{1}{6}\right)^2 \left(\frac{5}{6}\right)^3$$

$$= \frac{5!}{2!\;3!} \cdot \left(\frac{1}{6}\right)^2 \left(\frac{5}{6}\right)^3 = 0.16$$

(iii) $n = 5, r = 5$

$$\therefore \quad P(5) = \frac{5!}{5!\;0!} \cdot \left(\frac{1}{6}\right)^5 \left(\frac{5}{6}\right)^{5-5}$$

$$= \left(\frac{1}{6}\right)^5 = 1.286 \times 10^{-6}$$

(iv) The probability of number '4' not appearing uppermost in one dice $= \frac{5}{6}$. Therefore the probability that '4' does not appear uppermost in all five dice simultaneously $= \left(\frac{5}{6}\right)^5 = 0.4$

(v) The probability that the number '4' appears uppermost in atleast one dice.

$$= 1 - \left(\frac{5}{6}\right)^5$$

$$= 1 - 0.4 = 0.6$$

Problem 9 : A coin is so tossed that the probability of getting 'head' in a toss is 0.7. Deduce the probability that in 5 tosses, we get

(i) 2 heads 3 tails

(ii) all tails

(iii) all heads

[Agra 1980]

Solution : The probability of getting r heads and (n – r) tails with weight factor 'a' of getting head in a toss is given by

$$P[r_1(n-r)] = \frac{n!}{r!\,(n-r)!}.a^r.b^{n-r}$$

(i) $r = 2, n - r = 3$, Also $a = 0.7$ and $b = 1 - 0.7 = 0.3$ and $n = 5$.

$$\therefore\ P(2,3) = \frac{5!}{2!\ 3!}.(0.7)^2.(0.3)^5 = 0.1323$$

(ii) $r = 0, (n - r) = 5 - 0 = 5$

$$\therefore\ P(0,5) = \frac{5!}{0!\ 5!}.(0.7)^0.(0.3)^5 = 0.00243$$

(iii) $r = 5, (n - r) = 5 - 5 = 0$

$$\therefore\ P(5,0) = \frac{5!}{5!\ 0!}.(0.7)^5.(0.3)^0 = 0.16807$$

Problem 10 : Four distinguishable coins are tossed for a large number of times. Write down the different micorstates which may be observed and the macrostate into which they would fall. Write down the probability of the most probable macrostate.

Solution : Let us denote the coins by A, B, C, D. The macrostates and microstates of these coins, on being tossed for a large number of times will be as follows :

Microstate		Macrostate
Coins with head up	Coins with tails up	n_1 = number of heads up n_2 = number of tails up
A B C D	–	$n_1 = 4, n_2 = 0$

ABC	D	$n_1 = 3, n_2 = 1$
ABD	C	$n_1 = 3, n_2 = 1$
ACD	B	$n_1 = 3, n_2 = 1$
BCD	A	$n_1 = 3, n_2 = 1$
AB	CD	$n_1 = 2, n_2 = 1$
AC	BD	$n_1 = 2, n_2 = 2$
AD	BC	$n_1 = 2, n_2 = 2$
BC	AD	$n_1 = 2, n_2 = 2$
BD	AC	$n_1 = 2, n_2 = 2$
CD	AB	$n_1 = 2, n_2 = 2$
A	BCD	$n_1 = 2, n_2 = 2$
B	ACD	$n_1 = 1, n_2 = 3$
C	ABD	$n_1 = 1, n_2 = 3$
D	ABC	$n_1 = 1, n_2 = 3$
–	ABCD	$n_1 = 0, n_2 = 4$

Thus we find that number of microstates = 16

and that of macrostates = 5

The most probable macrostate is n = 2; n = 2 and its probability is

$\frac{6}{16}$ or $\frac{3}{18}$.

Problem 11 : In a random distribution of 10 particles between two boxes with equal probability, calculate the following :

(i) Probability of distribution (3, 7).

(ii) Ratio of probability of this distribution with the distribution of the largest probability.

(iii) The total number of microstates and the number of macrostates.

(iv) The number of microstates and the number of macrostates.

(v) The number of microstates in marcostate (3, 7).

[Agra 1971s]

Solution : For P (r, s), we have

$$P(r,s)=\frac{n!}{r!\ s!}\cdot\frac{1}{2^n} \qquad \text{Where } s = n - r$$

(i) $r = 3,\ s = 7,\ n = 10$

$$\therefore\ P(3,7)=\frac{10!}{3!\ 7!}\cdot\frac{1}{2^{10}}$$

$$=120\times 2^{-10}$$

(iii) For largest probability distribution, $r=\frac{n}{2}=\frac{10}{2}=5$

then $s = 10 - 5 = 5$

$\therefore$ largest probability, $P(5,5)=\frac{10!}{5!\ 5!}\cdot\frac{1}{2^{10}}=252\times 2^{-10}$

The ratio of P (3, 7) to that of the largest probability

$$=\frac{P(3,7)}{P(5,5)}=\frac{120\times 2^{-10}}{252\times 2^{-10}}=\frac{10}{21}$$

(iii) Total number of microstates $= 2^n = 2^{10}$

Total number of microstates = number of terms in binomial expansion

$(a + b)^n$ = number of binomial terms in $(a + b)^{10}$

$= 11$

The number of microstates in macrostate (3, 7)

$$=\frac{n!}{r!(n-r)!}=\frac{10!}{3!7!}=120$$

Problem 12 : Consider a system of three cells such that $n_1 = 6$, $n_2 = 4$, $n_3 = 3$, $E_1 = 0$, $E_2 = 2$. Joules per particle, $E_3 = 4$ joules per particle. If $\delta n_2 = -2$, find δn_1, and δn_2 assuming N and E to be constant. Also find the initial and final probabilities.

[Rohilkhand 1980s]

Solution : We have,

The total number of molecules, N

$$N = n_1 + n_2 + n_3 = 6 + 4 + 3 = 13$$

Also $E = E_1n_2 + E_2n_2 + E_3n_3$

$$= (0 \times 6) + (2 \times 4) + (4 \times 3) \text{Joules}$$

$$= 20 \text{ Joules}$$

Given : $\delta n_3 = -2$, so 2 molecules leave the third cell and go to first and second cells.

Since $n_1 + n_2 + n_3 = 13$

$$\therefore \delta n_1 + \delta n_2 + \delta n_3 = 0$$

or $\delta n_1 + \delta n_2 - 2 = 0$

or $\delta n_1 + \delta n_2 = 2$ . . . (i)

Also $E_1n_1 + E_2n_2 + E_3n_3 = 20$ Joules

$$0 \times n_1 + 2n_2 + 4n_3 = 20 \text{ Joules}$$

$$2n_2 + 4n_3 = 20 \text{ Joules}$$

or $2\delta n_2 + 4\delta n_3 = 0$

or $2\delta n_2 = -4\delta n_3 = -4 \times (-2) = 8$

$$\therefore \delta n_2 = \frac{8}{2} = 4$$

Substituting in equation (i), we get

$$\delta n_1 + 4 = 2$$

or $\delta n_1 = -2$

Hence the new distribution is

$$n_1' = n_1 + \delta n_1 = 6 - 4 = 4$$

$$n_2' = n_2 + \delta n_2 = 4 + 4 = 8$$

$$n_3' = n_3 + \delta n_3 = 3 - 2 =$$

The initial probability $= \dfrac{N!}{n_1!\ n_2!\ n_3!} = \dfrac{13!}{6!\ 4!\ 3!}$

$$\text{The final probability} = \frac{N!}{n_1'!\, n_2'!\, n_3'!} = \frac{13!}{4!\ 8!\ 1!}$$

Problem 13 : For the phase space representing a single particle of mass m in a volume V, calculate the number of phase cells in energy range 0 to E, given that volume of each phase cell is h^3. What will be the number of microstates in this energy range ?

[Rohilkhand 1990]

Solution : For any single particle in a volume V there are three position coordinates q_x, q_y, q_z and three momentum coordinates p_x, p_y, p_z to specify the microstates in phase space. An element of volume in phase space is

$$dp_x\, dp_y\, dp_z\, dq_x\, dq_y\, dq_z = dp_x\, dq_x\, dp_y\, dq_y\, dp_z\, dq_z = h^3$$

$[\quad h = dp_i dq_i]$

The total volume of the phase space will be

$$\iiint \iiint dp_x dp_y dp_z dq_x dq_y dq_z$$

where the limits of integration are to be specified we have

$$\iiint dq_x dq_y dq_z = V$$

∴ the volume of phase space will be

$$\iiint dp_x dp_y dp_z$$

∴ The number of phase cells

$$= \frac{\text{volume of phase space}}{\text{volume of one cell}}$$

$$= \frac{V \iiint dp_x dp_y dp_z}{h^3}$$

$$\text{But} \quad \iiint dp_x dp_y dp_z = \frac{4}{3}\pi P_{max}^3 = \frac{4\pi}{3}(2m.E)^{3/2}$$

$$\text{where} \quad P_{max}^2 = \left(p_x^2 + p_y^2 + p_z^2\right)_{max} = 2m.E$$

The number of cells in phase space,

$$\phi(E) = \frac{V}{3} \times \frac{4}{3}\pi(2mE)^{3/2}$$

$$= \frac{4\pi v}{3h^3}(2mE)^{3/2}$$

For any particle the number of accessible microstates will be equal to the number of accessible cells in the phase space, therefore the number of microstates in energy range 0 to E is given by

$$\phi(E) = \frac{4\pi V}{3h^3}(2mE)^{3/2}$$

Problem 14 : For a single particle of mass m enclosed in a volume v show that the number of accessible microstates in the energy range E to $E + \delta E$ is given by

$$\Omega(E) = \frac{4\sqrt{2}\pi v}{h^3} m^{3/2} E^{1/2} \delta E$$

where the phase space is divided by the rule $\delta p_i \delta q_i = h$

[Allahabad 1989]

Solution : The number of microstates in the phase space having energies between E and $E + \delta E$ is obtained by differentiating equations (2) with respect to E.

$$i.e.\ \Omega(E) = \frac{\partial\phi(E)}{\partial E} = \frac{4\pi v}{3h^3}(2mE)^{1/2}\, 2m\delta E.$$

$$= \frac{4\sqrt{2}\pi v}{h^3} m^{3/2} E^{1/2} \delta E.$$

Problem 15 : Use statistical definition of entropy to show that the difference in entropy between a state of volume V_f and a state of volume V_i (temperature and number of molecules remaining constant) is equal to nR log $\frac{V_f}{Vi}$

Solution : If V_i = volume in initial state

and V_f = volume in final state

Consider a rectangular box of volume V_f having a patition wall X that could be placed in any place with in the box so as to divide the box into two parts V_i and $V_f - V_i$. (See fig.)

Consider any single molecule of an ideal gas present in the box, then the probability of the molecule being found in volume V_i at any given instant in V_i/V_f.

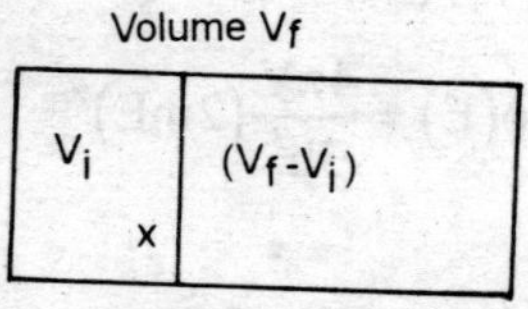

Fig.

In case two molecules are present in the box then probability of finding both the moelcules in volume V_i would be $\left(\frac{V_i}{V_f}\right)^2$

Similarly for P molecules the probability will be given by $\left(\frac{V_i}{V_f}\right)^p$

In n-gm moles of a perfect gas in present in the box then probability of finding all the molecules in volume V_i is

$$W_i = \left(\frac{V_i}{V_f}\right)^{n.N} \quad \text{Where N = Avagadro's number}$$

If the partition is removed, in such a case, then the probability of finding all the molecules in $V_f = 1$. i.e. $W_f = 1$.

Hence $$\frac{W_i}{W_f} = \left(\frac{V_i}{V_f}\right)^{n.N}$$

or $$\log = \frac{W_i}{W_f} = n.N \log \frac{V_i}{V_f}$$

Multiplying both sides by Bozmann's constant k,

we get

$$k \log = \frac{w_i}{w_f} = n\,k\,N \log \frac{v_i}{v_f}$$

or $$k \log = \frac{w_i}{w_f} = n\,R \log \frac{v_i}{v_f} \quad ...(i)[\because k.N = R]$$

Let S_i and S_f denote the entropy of the gas in volume V_i and V_f respectively. Using the statestical definition of entropy s = k log w, we get.

$$\Delta S = S_f - S_i = k \log w_f - k \log w_i$$

$$= K \log \frac{w_f}{w_i}$$

$$\therefore \Delta S = n\,R \log \frac{v_i}{v_f} \text{[using eq.(i)]}$$

This is the required expression. It gives the entorpy change in the expansion of a gas from volume v_i to v_f.

Problem 16 : Two states with energy difference 4.83×10^{-21}. Joules occur with relative probability e^2. Calculate the temperature. Given $k = 1.38 \times 10^{-23}\,J/k$ [Agra 1979]

Solution : Let E_1 and E_2 be the energy of two states the probability P_1 and P_2 of the states of energies E_1 and E_2 are given by

$$P_1 = Ce^{-E_1/kT}$$

$$P_2 = Ce^{-E_2/kT}$$

$$\therefore \frac{P_1}{P_2} = \frac{ce^{-E_1/kT}}{ce^{-E_1/kT}} = e^{(E_2 - E_1)/kT}$$

or $$\ln \frac{P_1}{P_2} = \frac{E_2 - E_1}{kT}$$

$$\therefore T = \frac{E_2 - E_1}{k \ln \frac{P_1}{P_2}}$$

Here $E_2 - E_1 = 4.83 \times 10^{-21}$ Joules; $\frac{P_1}{P_2} = e^2$

$$\therefore \quad T = \frac{4.83 \times 10^{-21}}{k \ln (e)^2} = \frac{4.83 \times 10^{-21}}{k \times 2}$$

$$= \frac{4.83 \times 10^{-21}}{2 \times 1.38 \times 10^{-23}} = 175k$$

Problem 17 : A system can take only three different energy states $E_1 = 0$, $E_2 = 1.38 \times 10^{-21}$ Joules and $E_3 = 2.76 \times 10^{-21}$ Joules. These three states can occur in 2, 5 and 4 different ways respectively. Find the probability that at temperature 100 k the system may be

(i) in one of the microstate of energy E_3

(ii) in the ground state E_1.

Solution : The following three energy values (E_1, E_2, E_3) give the probabilities with which the microstates can occur,

$$P_1 = Ce^{-E_1/kT} = Ce^0 = C$$

$$P_2 = Ce^{-E_2/kT} = Ce^{\frac{-1.38 \times 10^{-21}}{1.38 \times 10^{-23} \times 100}}$$

$$= Ce^{-1}$$

$$P_3 = ce^{-E_3/kT} = Ce^{\frac{-2.76 \times 10^{-21}}{1.38 \times 10^{-23} \times 100}}$$

$$= Ce^{-2}$$

Since 2, 5, 4 microstates can occur in the three energy states, the probabilities are

$$P_1 = 2C, P_2 = 5Ce^{-1} \text{ and } P_3 = 4Ce^{-2}$$

But

$$\Sigma P = 1$$

$$\therefore \quad P_1 + P_2 + P_3 = 1$$

$$\text{or } 2C + 5C\,e^{-1} + 4C\,e^{-2} = 1$$

$$\text{or } C(2 + 5e^{-1} + 4e^{-2}) = 1$$

$$\text{or } C = \frac{1}{2 + 5^{e-1} + 4^{e-1}} = \frac{1}{2 + 1.84 + 0.54} \quad [\because e = 2.72]$$

$$= \frac{1}{4.38}$$

(i) The probability for the system to be in one of the microstates of energy E_3 is

$$P_3 = 4Ce^{-2} = \frac{4C}{e^2}$$

$$= \frac{4}{4.38 \times (2.72)^2} \quad \left[\because C = \frac{1}{4.38}, e = 2.72\right]$$

$$= 0.12$$

(ii) The probability of the system to be in ground state is

$$P_1 = 2C = \frac{2}{4.38}$$

$$= 0.45$$

13

Kinetic Theory

Problem 1 : Calculate the root mean square velocity at N.T.P. for (i) air (ii) hydrogen (iii) oxygen.

Solution :

(i) **For Air**

$P = 76 \times 13.6 \times 981$ dynes per square centimetre

$\ell = 0.00129$ gm/cc at N.T.P.

$$\therefore \; c = \sqrt{\frac{3 \times 76 \times 13.6 \times 981}{0.00129}} \left[\because c = \sqrt{\frac{3p}{\varepsilon}} \right]$$

$$= 4.856 \times 10^4 \text{ cm/s}$$

(ii) **For Hydrogen**

$l = 0.000089$ gm/cc

$$\therefore \qquad c = \sqrt{\frac{3 \times 76 \times 13.6 \times 981}{0.000089}}$$

$$= 18.4 \times 104 \text{ cm/sec.}$$

(iii) **For Oxygen**

$\ell = 16 \times 0.000089$ g/cc.

[It has been assumed that density of oxygen is sixteen times the density of hydrogen]

$$\therefore\ c = \sqrt{\frac{3 \times 76 \times 13.6 \times 981}{16 \times 0.000089}}$$

$$= 4.6 \times 10^4 \text{ cm/sec.}$$

Problem 2 : The density of nitrogen at 0°C and 76 cm-Hg pressure is $0.000125 g/cm^3$. Find the root mean square velocity of the molecules.

Solution :

$$c = \sqrt{\frac{3 \times 76 \times 13.6 \times 981}{0.000125}}$$

$$= 4.93 \times 10^4 \text{ cm/s}$$

Problem 3 : Determine the root mean square velocity of hydrogen molecules at N.T.P.

(density of hydrogen at N.T.P. = 0.09 g/litre, density of mercury = 13.6 g/cc; g = 981 cm/sec^2)

Solution : The root mean square velocity, $\sqrt{c^2}$ is given by

$$\sqrt{c^2} = \sqrt{\frac{3p}{\ell}}$$

where P is in $dynes/cm^2$, ℓ is in gm/cc; $\sqrt{c^2}$ is cm/sec

Now $\ell = \frac{0.09}{1000} \text{gm/cc} = 0.00009 \text{gm/cc}.$

and P = 76 × 13.6 × 981 dynes/sq cm.

Since normal pressure = pressure at bottom of column of mercury 76 cm high

$$\therefore\ \sqrt{c^2} = \sqrt{\frac{3p}{\ell}} = \sqrt{\frac{3 \times 76 \times 13.6 \times 981}{0.00009}}$$

$$= 18.4 \times 10^4 \text{ cm/sec.}$$

Problem 4 : Calculate the root mean square velocity of molecules of a gas for which specific heat at constant pressure is 6.84 cal per gm mole per deg C. The velocity of the sound in the gas being 1300 meters per second. Gas constant R = 8.31 × 10^7 ergs per mole per deg C.

Solution :

The velocity of sound in a gas is given by $\sqrt{\frac{\gamma p}{\ell}}$, where γ is the ratio $\frac{Cp}{Cv}$ of the two specific heats of the gas, p is the pressure and ℓ is the density of the gas.

$$\therefore \sqrt{c^2} : v\sqrt{\frac{3\,p}{\ell}} : \sqrt{\frac{\gamma\,p}{\ell}} = \sqrt{3} : \sqrt{\gamma}$$

$$\sqrt{c^2} : 1300 = \sqrt{3} : \sqrt{\gamma} \qquad \ldots .(i)$$

Where $\sqrt{c^2}$ is in meters per second.

$$\text{Since } c_p - c_v = \frac{R}{J}$$

$$\therefore \; 6.84 - c_v = \frac{8.31 \times 10^7}{4.18 \times 10^7} = 1.99 \quad \left[\because \; J = 4.18 \times 10^7 \text{ ergs for colorie}\right]$$

$$c_v = 6.84 - 1.99$$

$$= 4.85 \text{ cal per gm per °C}$$

$$\therefore \qquad \gamma = \frac{c_p}{c_v} = \frac{6.84}{4.85} = 1.4$$

Substituting in equation (i), we get

$$\sqrt{c^2} : 1300 = \sqrt{3} : \sqrt{1.41}$$

$$\therefore \qquad \sqrt{c^2} = \frac{1300 = \sqrt{3}}{\sqrt{1.41}}$$

$$= 1898 \text{ meters per second}$$

Problem 5 : The best vacuum attainable so far in the laboratory is 10^{-10} mm of Hg. Calculate the number of molecules per cc at 27°C in this vacuum. Given Avagadro's number N = 6 × 10^{23}/gm mole.

[Agra 1976]

Solution : For one mole of an ideal gas, we have

$$PV = RT. \qquad \text{[Gas equation]}$$

Now at N.T.P.,

$$P = 76 \text{ cm of Hg} = 76 \times 13.6 \times 980 \text{ dynes/cm}^2$$

$$V = 22.4 \text{ litre} = 24000 \text{ cc}$$

$$T = 0°C = 273k.$$

In the given conditions

$$P_1 = 10^{-10} \text{ mm of Hg}$$

$$= 10^{-11} \text{ cm of Hg} = 10^{-11} \times 13.6 \times 980 \text{ dynes/cm}^2$$

$$T_1 = 27°C = 300 \text{ k}$$

$$v_1 = ?$$

Using gas equation, we get

$$v_1 = \frac{RT_1}{P_1}$$

$$= \frac{Pv}{T} \cdot \frac{T_1}{P_1} \qquad \left[\because R = \frac{Pv}{T}\right]$$

or $$v_1 = \frac{76 \times 13.6 \times 980 \times 22400 \times 300}{273 \times 10^{-11} \times 13.6 \times 980}$$

$$= \frac{76 \times 22400 \times 300}{273 \times 10^{-11}}$$

$$= 1.87 \times 10^{16} \text{ cc.}$$

$$\text{Number of molecules/ cc} = \frac{6 \times 10^{23}}{10^{-11} \times 273}$$

$$= 3.20 \times 10^{7}$$

Problem 6 : Explain on the basis of kinetic theory, why upper atmosphere is cooled while its kinetic temperature is of the order of 1000k.

[Gorakhpur 1978]

Solution : The heat content of an individual molecule in the upper atmosphere is about 3.5 times as compared to that on the surface of the earth. On the surface of the earth the kinetic temperature is about 300 k where as in the upper atmpoɔphere it is about 1000k. However, the pressure and so the density is very low in the upper atmosphere. Thus in the upper atmosphere the number of molecules per cc is very small. Thus obviously the total heat content for the same volume of the gas, in the upper atmosphere is very small in comparison to the heat content on the surface of the earth. Because of this the temperture of the upper atmosphere is low.

Problem 7 : Find the value of v_x for which the probability falls to (i) $\frac{1}{e}$ times (ii) $\frac{1}{10}$ times, the maximum value.

Solution : The probability of a molecule having x component of velocity between v_x to $v_x + dv_x$

$$P(v_x) = \left(\frac{m}{2\pi kT}\right)^{1/2} e^{-mv_x^2/2kT} \qquad \ldots \text{(i)}$$

The probability will be maximum when $v_x = 0$, i.e.,

$$P_{max}(v_x) = \left(\frac{m}{2\pi kT}\right)^{1/2}$$

(i) Given $P(v_x) = \frac{1}{e}(P_{max}) = \frac{1}{e}\left(\frac{m}{2\pi kT}\right)^{1/2}$

$\therefore$ Substituting this value of $P(v_x)$ in (1), we get

$$\frac{1}{e}\left(\frac{m}{2\pi kT}\right)^{1/2} = \left(\frac{m}{2\pi kT}\right)^{1/2} . e^{-mv_x^2/2kT}$$

or $$\frac{1}{e} = e^{-mv_x^2/2kT}$$

taking log, we get

$$\log_e e = -\frac{mv_x^2}{2kT}$$

$$\therefore\ v_x = \sqrt{\left(\frac{2kT}{m}\right)}$$

(ii) Given : $P(v_x) = \frac{1}{10} P_{max.} = \frac{1}{10}\left(\frac{m}{2\pi kT}\right)^{1/2}$

or $$\frac{1}{10}\left(\frac{m}{2\pi kT}\right)^{1/2} = \left(\frac{m}{2\pi kT}\right) e^{-mv_x^2/2kT}$$

or $$\frac{1}{10} = e^{-mv_x^2/2kT}$$

taking log, we get

$$-\log_e 10 = \frac{-mv_x^2}{2kT}$$

or $$v_x = \sqrt{\left(\frac{2kT}{m}\log_e 10\right)} = \sqrt{\left(\frac{2kT}{m}\times 2.3\right)}$$

$$\therefore\ v_x = \sqrt{\left(\frac{4.6\ kT}{m}\right)}$$

Problem 8 : Let v_x, v_y, v_z represent the three Cartesian components of velocity of a molecule in a gas. Using symmetry considerations and equipartition thereon, deduce expressions for the following mean values in terms of k, T and m :

(i) $\langle v_x \rangle$, (ii) $\langle {v_x}^2 \rangle$,

(iii) $\langle v_x v_z \rangle$, (iv) $\langle (v_x + bv_y)^2 \rangle$,

where be is a constant. (Agra 1975)

Solution :

(i) We have

$$\bar{v}_x = \langle v_x \rangle = \int_{-\infty}^{+\infty} v_x \left(\tfrac{m}{2\pi kT}\right)^{1/2} e^{-mv_x^2/2kT} dv_x$$

$$= \left(\frac{m}{2\pi kT}\right)^{1/2} \int_{-\infty}^{+\infty} v_x e^{-mv_x^2/2kT} dv_x$$

$$= 0 \left(\text{since} \int_{-\infty}^{+\infty} v_x e^{-mv_x^2/2kT} dv_x = 0\right)$$

(ii) $$\overline{v_x^2} = \left\langle v_x^2 \right\rangle = \int_{-\infty}^{+\infty} v_x P(v_x)\, dv_x$$

$$= \int_{-\infty}^{+\infty} \left(\frac{m}{2\pi kT}\right)^{1/2} v_x^2 e^{-mv_x^2/2kT}\, dv_x$$

$$= \left(\frac{m}{2\pi kT}\right)^{1/2} = \int_{-\infty}^{+\infty} v_x^2 e^{-mv_x^2/2kT}\, dv_x$$

$$\left(\frac{m}{2\pi kT}\right)^{1/2} \frac{1}{2} \sqrt{\left\{\frac{\pi}{(m/2kT)^3}\right\}}$$

$$\left[\text{since} \int_{-\infty}^{+\infty} v_x e^{-mv_x^2/2kT}\, dv_x = \frac{1}{2} \sqrt{\left\{\frac{\pi}{(m/2kT)^3}\right\}}\right]$$

$$= \frac{kT}{m}.$$

(iii) $$\overline{v_x . v_z} = \langle V_x - V_z \rangle$$

$$= \int_{-\infty}^{+\infty} \int_{-\infty}^{+\infty} v_x\, v_z\, P\,(v_x\, v_z)\, dv_x\, dv_z$$

$$= \int_{-\infty}^{+\infty} \int_{-\infty}^{+\infty} v_x\, v_z\, P(v_x)\, P(v_z)\, dv_x\, dv_z$$

$$= \int_{-\infty}^{+\infty} v_x P(v_x)\, dv_x \int_{-\infty}^{+\infty} v_z\, P(v_z)\, dv_z$$

$$= \int_{-\infty}^{+\infty} v_x \left(\frac{m}{2\pi kT}\right)^{1/2} e^{-mv_x^2/2kT}\, dv_x$$

$$= \int_{-\infty}^{+\infty} v_z \left(\frac{m}{2\pi kT}\right)^{1/2} e^{-mv_z^2/2kT}\, dv_z$$

$$= \left(\frac{m}{2\pi kT}\right) \int_{-\infty}^{+\infty} v_x\, e^{-mv_x^2/2kT}\, dv_x$$

$$= \int_{-\infty}^{+\infty} v_z\, e^{-mv_z^2/2kT}\, dv_z$$

$$= 0$$

(iv) $\overline{\left(v_x + bv_y\right)^2} = \left\langle \left(v_x + bv_y\right)^2 \right\rangle$

$$\left\langle = \left(v_x{}^2 + 2bv_xv_y + b^2v_y{}^2\right)^2 \right\rangle$$

$$\left\langle v_x{}^2 \right\rangle + 2b\left\langle v_xv_y \right\rangle + b^2\left\langle v_y{}^2 \right\rangle$$

$$= \frac{kT}{m} + 0 + b^2\frac{kT}{m}$$

$$= \left(1 + b^2\right)\frac{kT}{m}.$$

Problem 9 : Calculate the probability that the speed of oxygen molecule lies betweeen 99.5 and 100.5 metres/sec, at 200 k.

Solution : The probability of any molecule to lie between speeds *v* and (*v* + dv) is given by

$$p_v = 4\pi\left(\frac{m}{2\pi kT}\right)^{3/2}.v^2.e^{\frac{-mv^2}{2kT}}.dv$$

Given : m = 32 amu. (For oxygen m = 32 amu)

$$= \frac{32}{6 \times 10^{26}} \text{ kg.}$$

$$v = 100\,\text{m/s}$$

$$dv = 100.5 - 99.5 = \text{metre/sec}$$

$$K = 1.38 \times 10^{-23}\ \text{J/deg k}$$

$$T = 200\text{k.}$$

$$P_v = 4 \times 3.14 \times \left[\frac{\frac{32}{6 \times 10^{26}}}{2 \times 3.14 \times 1.38 \times 10^{-23} \times 200}\right]^{3/2} \times 100^2$$

$$\times\left[e^{-\frac{32 \times 100^2}{2 \times 1.38 \times 10^{-23} \times 200}}\right] \times 1$$

$$= \frac{4\times 1}{\sqrt{3.14}}\left[\frac{32}{2\times 1.38\times 10^{-23}\times 200\times 6\times 10^{26}}\right]^{3/2}\times 10^{4}$$

$$\times\left[e^{-\frac{32}{2\times 1.38\times 10^{-23}\times 200\times 6\times 10^{26}}}\right]^{3/2}$$

$$= \frac{4}{1.77}\left(0.96\times 10^{-5}\right)^{3/2}\times 10^{4}\times e^{-9.6\times 10^{-2}}$$

$$= 6.11\times 10^{-4}$$

or $P_v = 6.11\times 10^{-4}$

Problem 10 : At what temperature will the root mean square velocity of hydrogen be doubled of its value at N.T.P., pressure remaining constant.

Solution :

We know $C \alpha \sqrt{T}$

Let the mean square velocity (c) at N.T.P. (i.e. at T_0) be C_0.

If T is the required temperature, the velocity at this temperature is $2C_0$.

$$\therefore \quad \frac{c}{c_0} = \frac{2c_0}{c_0} = \sqrt{\frac{T}{T_0}}$$

$$\text{or} \quad 4 = \frac{T}{T_0}$$

$$\text{or} \quad T = 4\times T_0$$

$$= 4\times 273$$

$$= 1092\text{k} \text{ or } 819°\text{C}$$

Problem 11 : Calculate the root mean square velocity of air at (i) –100°C (ii) 1000°C. (Given the density of air at N.T.P = 0.001296 gm/cm^3)

Solution : The R.M.S. velocity at any temperature T°C is given by

$$c = c_6 \sqrt{\frac{273 + T}{273}}$$

$$= \sqrt{\frac{3P\,(273 + T)}{\ell\,.273}} \qquad \left[\because c_0 = \sqrt{\frac{3p}{\ell}}\right]$$

(i) $\ell = 0.001296$, $T = -100°C$

$$\therefore \chi = \sqrt{\frac{3 \times 76 \times 13.6 \times 980 \times (273 - 100)}{0.001296 \times 273}}$$

$$= 1.48 \times 100^4 \text{ cm / sec}$$

(ii) $T = 1000°C$

$$\therefore C = \sqrt{\frac{3 \times 76 \times 13.6 \times 980 \times (273 + 1000)}{0.001296 \times 273}}$$

$$= 10.87 \times 100^4 \text{ cm / sec}$$

Problem 12 : You are given the following group of particles (N_i represents the number of particles with speed V_i)

N_i	V_i *(m/s)*
2	1.0
4	2.0
8	3.0
6	4.0
3	5.0

(i) Compute the average speed $\langle v \rangle$

(ii) Compute the r.m.s. (V v.m.s.)

(iii) Among the five speeds shown, which is most probable speed for the entire group.

(Agra 1971)

Solution :

(i) The average speed $\langle v \rangle$ is given by

$$\langle v \rangle = \frac{\Sigma N_i v_i}{\Sigma N_i}$$

$$= \frac{2 \times 1 + 4 \times 2 + 8 \times 3 + 6 \times 4 + 3 \times 5}{2 + 4 + 8 + 6 + 3}$$

$$= \frac{73}{23} = 3.17 \text{ m/s}$$

(ii)

$$V_{r.m.s.} = \sqrt{\frac{\Sigma N_i v_i^2}{\Sigma N_i}}$$

$$= \sqrt{\frac{2 \times 1^2 + 4 \times 2^2 + 8 \times 3^2 + 6 \times 4^2 + 3 \times 5^2}{2 + 4 + 8 + 6 + 3}}$$

$$= \sqrt{\frac{2 + 16 + 72 + 96 + 75}{23}} = \sqrt{\frac{261}{23}}$$

$$= 3.36 \text{m/s}$$

(iii) out of a total of 23 particles, 8 particles have a speed of 3m/s, 6 particles have a speed of 4 m/s, 4 particles have a speed of 2 m/s and the remaining 5 particles have different speeds. Thus the most probable speed is 3 m/s.

Problem 13 : Calculate the root mean square speed of a molecule of hydrogen at N.T.P. the Boltzmann constant is 1.38×10^{-23} Joule per degree and Avagadro's number is 6×10^{26} (kg mole^{-1}).

Solution : The root mean square speed V.v.m.s. is

$$V_{r.m.s.} = \sqrt{\frac{3kT}{m}}$$

Given : $K = 1.38 \times 10^{-23}$ J / degree; T = 273k

$$m = \frac{2}{6} \times 10^{-26} \text{kg} \qquad [\because \text{ For hydrogen M } = 2]$$

Thus $V_{r.m.s.} = \sqrt{\dfrac{3KT}{m}} = \sqrt{\dfrac{3 \times 1.38 \times 10^{-23} \times 273}{\frac{2}{6 \times 10^{26}}}}$

$= 1838 m/s$

Problem 14 : At what temperature, pressure remaining constant, will the molecular velocity (root mean square velocity) of hydrogen will be double of its value at N.T.P.

[Gorakhpur 1983]

Solution : We know

$$V_{r.m.s.} = \sqrt{\frac{3KT}{m}}$$

So $V_{r.m.s.}$ of hydrogen at N.T.P. is

$$V_{r.m.s.} = v_1 = \sqrt{\frac{3k \times 273}{m}} \quad \ldots \text{(i)} \quad [\because T = 273\text{ k}]$$

At temp. T_1 $V_{r.m.s.} = v_2 = \sqrt{\dfrac{3KT}{m}}$ . . . (ii)

But $v_2 = 2\, v_1$

∴ From (i) and (ii) we get

$$2 = \sqrt{\frac{T}{273}}$$

or $4 = \dfrac{T}{273}$

or $T = 273 \times 4 = 1092$ k or 819°C.

Problem 15 : If the root mean square value of the molecules of hydrogen at N.T.P. is 1.84 km/s. Calculate the root mean square velocity of the oxygen molecules at N.T.P. Molecular weight of hydrogen and oxygen are 2 and 32 respectively.

(Gorakhpur 1983)

Solution : Let v_1 and v_2 be the values of root mean square velocity of hydrogen and oxygen molecules respectivley,

Then $$v_1 = \sqrt{\frac{3KT}{2}},\ v_2 = \sqrt{\frac{3KT}{32}}$$

$$\therefore \frac{v_2}{v_1} = \sqrt{\frac{2}{32}} = \sqrt{\frac{1}{16}}$$

Since $v_1 = 1.84$ km/s (given)

$$\therefore \quad v_2 = 1.84 \times \sqrt{\frac{1}{16}}$$

$$= \frac{1.84}{4} = 0.46\text{km/s}$$

Problem 16 : Calculate the temperature at which the root mean square velocity is equal to the escape velocity from the surface of earth (i) For hydrogen (ii) For oxygen

Solution : K.E. of a molecule $= \frac{1}{2}mu^2$

where U is the root mean square velocity.

Since $$u^2 = \frac{3\ kT}{m}$$

$$\therefore \frac{1}{2}mu^2 = \frac{3}{2}kT$$

Escape velocity, $v_e = \sqrt{2gR}$, where R = radius of earth

Since $u = v_e$ (given)

$$\therefore \text{ K.E. of molecule} = \frac{1}{2}mu^2 = \frac{1}{2}m\,v_e^2$$

$$= \frac{1}{2}m \times 2gR = mgR$$

or $$T = \frac{2}{3}\frac{m\,g\,R}{k}$$

(i) *Temperature for hydrogen*

$$T = \frac{2}{3}\frac{m_H .\ g.\ R.}{k}$$

$$= \frac{2}{3} \times \frac{\left(\frac{2}{6} \times 10^{-26}\right) \times 9.8 \times 6.4 \times 10^6}{1.38 \times 10^{26}}$$

$$= 1.0 \times 10^4 k.$$

(ii) *Temperature for oxygen*

$$T = \frac{2}{3}\ \frac{m_0 .g.R}{k} = \frac{2}{3} \times \left(\frac{32}{6 \times 10^{26}}\right) \times \frac{9.8 \times 6.4 \times 10^6}{1.38 \times 10^{-23}}$$

$$= 16 \times 10^4 \text{ k.}$$

Problem 17 : The first excited state of hydrogen atom is 10.2 ev above its ground state. What temperature is needed to excite hydrogen atoms to the first excited level?

[Gorakhpur 1985]

Solution : At any temperature Tk, we have

$$\frac{1}{2} mu^2 = \frac{3}{2} kT$$

Given : $\frac{1}{2} mu^2 = 10.2 ev = 10.2 \times 1.6 \times 10^{-19} J$

$$\frac{3}{2} kT = 10.2 \times 1.6 \times 10^{-19} J$$

$$\text{or } T = \frac{3}{2} \times \frac{10.2 \times 1.6 \times 10^{-19}}{3 \times 1.38 \times 10^{-23}}$$

$$\left[\therefore k = 1.38 \times 10^{-23} J / k\right]$$

$$= 7.88 \times 10^4 k. \qquad \left[\because k = 1.38 \times 10^{-23}\ J / k\right]$$

Problem 18 : Calculate the root mean square speed and most probable speed of a gas whose density is 1.4 gm/litre at a pressure of $10^5 N/m^2$.

[Gorakhpur 1985]

Solution :

$$P = 10^5\ N/m^2,\ \ell = 1.4\ g/\ell = 1.4\ kg/m^3$$

$$u = \sqrt{\frac{3P}{\ell}} = \sqrt{\frac{3\times 10^5}{1.4}} = 4.6\times 10^2\ m/s$$

Most probable speed, $\alpha = \sqrt{\frac{2}{3}} \times 4$

$$= \sqrt{\frac{2}{3}} \times 4.6 \times 10^2$$

$$= 3.77 \times 10^2\ m/s$$

Problem 19 : At what temperature will the average speed of hydrogen molecules be the same as that of nitrogen molecules kept at 35°C. Molecular weight of nitrogen and hydrogen are 2.8 and 2 atomic mass units respectively. Also calculate the most probable speed of nitrogen molecules at this temperature. (Given Boltzmann's constant, $k = 1.38 \times 10^{-23}$ Joules per degree).

(Rajasthan 1986)

Solution : If V_1 and V_2 are the average speeds of hydrogen and nitrogen molecules respectively at temperature T_1 k and T_2 k, Then

$$V_1 = \sqrt{\frac{8\,k\,T_1}{m_1\pi}} \quad \text{and} \quad V_2 = \sqrt{\frac{8\,k\,T_1}{m_2\pi}}$$

where m_1 = mass of hydrogen molecule,

m_2 = mass of nitrogen moelcule.

From the above equation, we get

$$\frac{V_1}{V_2} = \sqrt{\frac{T_1}{m_1} \Big/ \frac{T_2}{m_2}} = \sqrt{\frac{T_1\,m_2}{T_2\,m_1}}$$

Given : $T_2 = 35°C = 308\ K$

$V_1 = V_2$

$$\frac{m_2}{m_1} = \frac{28}{2} = 14$$

Substituting these values in the above equation,

$$1 = \sqrt{\frac{T_1 \times 14}{308}}$$

or $$T_1 = \frac{308}{14} = 22k.$$

The most probable speed of nitrogen at temperature Tk is given by

most probable speed, $\alpha = \sqrt{\frac{2\ \hbar T}{m}}$

Given : $k = 1.38 \times 10^{-23}\ J / degree, T = 22K.$

$$m = \frac{.28}{6 \times 10^{26}} Kg$$

$$\therefore\ \alpha = \sqrt{\frac{2 \times 1.38 \times 10^{-23} \times 22}{\frac{28}{6 \times 10^{26}}}}$$

$$= \sqrt{\frac{2 \times 1.38 \times 10^{-23} \times 22 \times 6 \times 10^{26}}{28}}$$

$$= 1.14 \times 10^{2}\ m/s$$

Problem 20 : Calculate the kinetic energy of each molecule of an ideal monoatomic gas at N.T.P. (Boltzmann constant, $k = 1.38 \times 10^{-23}$ J/degree)

Solution : Each molecule of an ideal monoatomic gas can have only three degrees of freedom since it can have only translatory motion. In accordance with the law of equipartition of energy, the energy associated with each degree of freedom $3 \times \frac{1}{2} kT = \frac{3}{2} kT$

$\therefore$ kinetic energy of each molecule of an ideal monoatomic gas $= \dfrac{3kT}{2}$

Given : $k = 1.38 \times 10^{-23}$ Joules/degree K

$T = 27$ K

$$\therefore \text{ K.E. of each molecule} = \frac{3}{2} \cdot \times 1.38 \times 10^{-23} \times 273\text{J}$$

$$= 5.65 \times 10^{-21}\text{J}$$

Problem 21 : Find the molecular kinetic energy of 1 g of helium at N.T.P. what will be the energy at 100°C?

Solution : Using the relation

$$\frac{1}{2} C^2 = \frac{3}{2} \cdot \frac{R}{M} . T$$

We can find the K.E. at N.T.P.

$R = 8.3 \times 10^7$; $T = 273$ K; $M = 4$ (for helium)

$$\therefore \text{ K.E. } \frac{1}{2} \times 1 \times C^2 = \frac{3}{2} \times \frac{8.3 \times 10^7 \times 273}{4} = 8.5 \times 10^9 \text{ ergs.}$$

To find energy at 100°C

$$\text{K.E. } \frac{8.5 \times 10^9 \times 373}{273} \qquad \left[\because C = C_0 \sqrt{\frac{273+T}{273}}\right]$$

$$= 1.16 \times 10^{10} \text{ ergs.}$$

Problem 22 : Calculate the root mean square speed of a nitrogen molecule at 0°C.

Solution : We know that average K.E. associated with one molecule is

$$\frac{3}{2} \cdot \frac{R}{N} . t = \frac{3}{2} kT \qquad \left[\text{where } k = \frac{R}{N}\right]$$

Thus

$$\therefore V_{r.m.s.} = \sqrt{\frac{3kT}{m}} \qquad \text{[Where m = mass of one molecule]}$$

$$= \sqrt{\frac{3RT}{m}} \qquad \text{[M = Mol. wt of the gas]}$$

Thus

$$V_{r.m.s.} = \sqrt{\frac{3 \times 8.314 \times 273 kgm^2 s^{-1}}{20/1000 kg}} \qquad [\text{mol.wt. of } N_2 = 28g]$$

$$= 493 ms^{-1}$$

Problem 23 : The kinetic energy of a molecule of hydrogen at 0°C is 5.64×10^{-14} ergs and the molecular constant $R = 8.32 \times 10^7$ ergs. Calculate thae Avagadro's number, N.

Solution : K.E of a molecule $= \frac{1}{2} mC^2 = 5.64 \times 10^{-14}$ ergs

Using the relation $\frac{1}{3} N.m.C^2 = RT$, We get

$$N = \frac{3RT}{mC^2} = \frac{3 \times 8.32 \times 10^7 \times 273}{2 \times 5.64 \times 10^{-14}}$$

$$= 6.08 \times 10^{23}$$

Problem 24 : A container having adiabatic walls is divided into two parts with the help of a diathermic partition, and is kept horizontal (see fig). Both compartments are filled with an ideal gas. Volume, pressure and temperature in the left compartment are V, P and T respectively, while these parameters for the gas is the right compartment are 2V, 2P and T. The partition is now let go, and it moves without friction.

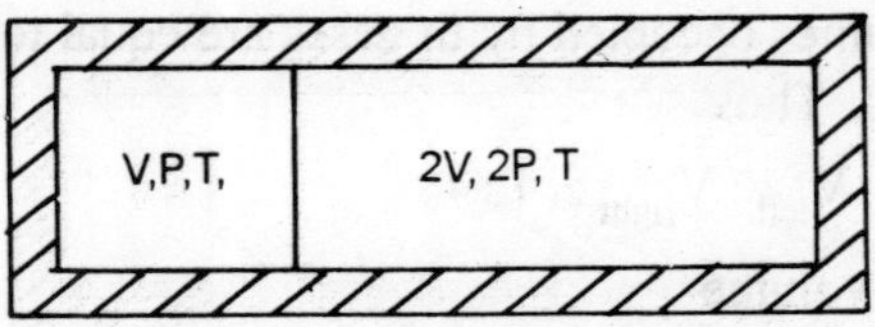

Fig.

(i) What are the relative amounts of the gas in the two compartments?

(ii) What are the tempertures on both sides?

(iii) What are the pressures on both sides?

(iv) What are the final volumes ?

Solution :

(i) *For the left compartment,*

$$N = \frac{\text{Pressure} \times \text{Volume}}{\text{Temperature}} = \frac{P \times V}{T}$$

For the right Compartment,

$$\frac{\text{Pressure} \times \text{Volume}}{\text{Temperature}} = \frac{4\ PV}{T}$$

Therefore, the amount of gas is the right hand compartment is four times greater than that in the left hand compartment.

(ii) When the partition is let to go, it will move to the left. Therefore the gas on the left of it will temporarily get heated up, and that on right will get cooled. After some time both temperatures will become equal.

If we take whole of the gas as one system, we have neither given any heat to it, nor any work has been done. Therefore, according to first law of thermodynamics, its internal energy will remain unchanged. But for an ideal gas internal energy is a function of temperature only, hence the final temperature will be the same as before i.e. T.K.

(iii) As the piston is frictionless, pressure in both compartments will be equal.

(iv) Since the T and P in the two compartments is same so the ratio of volumes occupied by th gases are equal to the ratio of their masses. Thus

$$V_{left} : V_{right} :: 1 : 4$$

But the total volume

$$V_{left} + V_{right} = 3V$$

Therefore volume of left compartment

$$V_{left} = \frac{3V}{5}$$

and $\quad V_{right} = \frac{12V}{5}$

Problem 25 : The mass of an oxygen molecule is 5.28×10^{-26} kg and its mean or average velocity $\langle v \rangle$ at N.T.P. is 4.25×10^2 m/s. Calculate the average K.E. of an oxygen molecule at 0°C.

(Agra 1975)

Solution : Average speed, $\langle v \rangle = \sqrt{\frac{8\,k\,T}{m\pi}}$

Root mean square velocity, $V_{r.m.s.} = \sqrt{\frac{3k\,T}{m}}$

$$\frac{V_{r.m.s.}}{\langle v \rangle} = \frac{\sqrt{3kT/m}}{\sqrt{\frac{8kT}{m\pi}}} = \sqrt{\frac{3\pi}{8}}$$

$$(V_{r.m.s.})^2 = \frac{3\pi}{8}(\langle v \rangle)^2$$

Thus average kinetic energy of oxygem

molecule $\quad = \frac{1}{2} m V_{r.m.s.}^2$

$$= \frac{1}{2} m . \frac{3\pi}{8}(\langle v \rangle)^2$$

$$= \frac{3}{16} m\pi(\langle v \rangle)^2$$

$$= \frac{3}{16} 5.28 \times 10^{-26} \times 3.14 \times 4.25 \times 10^2$$

$$= 5.614 \times 10^{-21}\,J$$

Problem 26 : Calculate average translational energy of a mercury atom at 27°C. Is it enough to ionise a mercuryatom during a collision ?

Solution : Average translational energy $= \frac{3}{2} kT$

$$= \frac{3}{2} \times 1.381 \times 10^{-23}\,JK^{-1} \times 300K$$

$$= 6.215 \times 10^{-21}\,J.$$

$$= \frac{6.215 \times 10^{-21}}{1.602 \times 10^{-19}}\,\text{electron volt}$$

$$= 0.0388\ eV$$

Since the I.E. of mercury is 10.4 eV so a mercury atom can not be ionised during a collision.

The excitation energy for a mercury atom is 4.9 eV so with this energy (0.0388 eV) the mercury atom can not even be excited.

Problem 27 : The first allowed excited state of hydrogen atom is 10.2 eV above its ground state. To what temperature should hydrogen gas be raised so that inelastic collision may excite an appreciable number of atoms to their first excited state ?

Solution : Let the required temperature be TK.

Then

$$\frac{3}{2}kT = 10.2 \times 1.602 \times 10^{-19}\,J \quad \left(\because 1ev = 1.602 \times 10^{-19}\,J\right)$$

$$\text{or } T = \frac{10.2 \times 1.602 \times 10^{-19}\,J}{1.381 \times 10^{-23}\,JK^{-1}} \times \frac{2}{3} \quad \left(\therefore k = 1.381 \times 10^{-23}\,J\right)$$

$$= 7.88 \times 10^{4}\,K\cdot$$

Problem 28 : Calculate the kinetic energy of translation of molecule of nitrogen gas at STP moving with the most probable speed.

Solution : We know that

Most probable velocity, $V_m = 0.817\ V_{r.m.s.}$

and average velocity, $V_{a.v.e.} = 0.921\ V_{r.m.s.}$

Where $V_{r.m.s.}$ is the root mean square velocity.

For 1 gm-mol of nitrogen,

$$V_{r.m.s.} = \sqrt{\frac{3PV}{M}} = \sqrt{\frac{3 \times 1.013 \times 10^5 Nm^{-2} \times 22.4 \times 10^{-3} m^3 mol^{-1}}{28 \times 10^{-3} kg\ mol^{-1}}}$$

$= 493 ms^{-1}$

Hence $V_m = 0.817 \times 493$

$= 403\ ms^{-1}$

Mass of one nitrogen molecule,

$$m = \frac{28}{6.02 \times 10^{23}} g$$

$$= 4.65 \times 10^{-26} Kg$$

(E_m) K.E. associated with a nitrogen molecule moving with most probable speed is

$$E_m = \frac{1}{2} \times 4.65 \times 10^{-26} kg \times \left(403\ ms^{-1}\right)^2$$

$$= 3.78 \times 10^{-21} J$$

$$= \frac{3.78 \times 10^{-21}}{1.60 \times 10^{-19}} eV$$

$$= 0.0236\ eV$$

Problem 29 : What is the kinetic temperature of an electron having energy 1 KeV?

Solution : Let the temperature be TK.

The K.E. of the electron moving with r.m.s. speed in an electorn gas at TK is

$$= \frac{1}{2} m_e V_{r.m.s.}^2 = \frac{3}{2} kT \qquad \text{[assuming it as a perfect gas]}$$

So the K.E. of an electron moving with most probable speed is

$$E_m = \frac{1}{2} m_e V_m^2 = \frac{1}{2} m_e \left(\sqrt{\tfrac{2}{3}} V_{r.m.s.}\right)^2$$

$$E_m = \frac{2}{3} \times \frac{1}{2} m_e V_{r.m.s.}^2$$

$$E_m = \frac{2}{3} \times \frac{1}{2} kT$$

$$E_m = kT$$

$$\therefore \; kT = 1 \text{ KeV} \quad \text{or}$$

$$\therefore \; T = \frac{1 \times 10^3 \times 1.6 \times 10^{-19} \text{ J}}{1.38 \times 10^{-23} \text{ JK}^{-1}}$$

$$= 1.16 \times 10^7 \text{ K}$$

$$= 11.6 \times 10^6 \text{ K or } 11.6 \text{ million K.}$$

Problem 30 : Find the mean free path, frequency of collision and the molecular diameter of nitrogen gas. Given $\eta = 166 \times 10^{-6}$ dyne/cm^2/unit velocity gradient, $\overline{C} = 4.5 \times 10^4$ cm/sec and $\ell = 1.25 \times 10^{-3}$ g/cm^3.

Solution : We know that $\eta = \frac{1}{3} \ell . \bar{c} . \lambda.$

where η = coefficient of viscosity,

ℓ = density

$\bar{c}$ = average velocity

λ = mean free path.

From the above relation, we get

$$\lambda = \frac{3\eta}{\ell . \bar{c}},$$

Substituting various values, we get

$$\lambda = \frac{3 \times 166 \times 10^{-6}}{1.25 \times 10^{-3} \times 4.5 \times 10^4}$$

$$= 9 \times 10^{-6} \text{ cm.}$$

$$\text{Frequencey of collisions} = \frac{\bar{c}}{\lambda} = \frac{4.5 \times 10^4}{9 \times 10^{-6}} = 5 \times 10^9$$

Assuming n = 2.7×10^{19} per cc. using Maxwell's relation

$$\lambda = \frac{1}{\sqrt{2}\,\pi\,\sigma^2 n}$$

$$\sigma = \left(\frac{1}{\sqrt{2}\,\pi\,n\,\lambda}\right)^{1/2}$$

$$= \left(\frac{1}{\sqrt{2} \times \frac{22}{7} \times 2.7 \times 10^{19} \times 9 \times 10^{-6}}\right)^{1/2}$$

$$= 3 \times 10^{-8}\ \text{cm.}$$

From these values we can get an idea of the magnitude of various quantities involved in kinetic theory of gases.

Problem 31 : The mean free path of molecules of a certain gas at pressure P and temperature T is 2×10^{-5} cm. Deduce the mean free path under the following conditions :

(i) Pressure $P \times 10^{-6}$, temperature T

(ii) Pressure $\frac{p}{2}$, temperature 2T

Also calculate the probability that under conditons, pressure P, temperature T, molecules would traverse 6×10^{-5} cm without collision.

(Agra 1972)

Solution : For mean free path at pressure P and temp. T, we have

$$\lambda = \frac{kT}{\sqrt{2}\left(\pi\,\sigma^2\,\ell\right)} \qquad \text{...(i)}$$

(i) Mean free path at pressure = $P \times 10^{-6}$ and temperature = T

$$\lambda_1 = \frac{kT}{\sqrt{2}\left(\pi\,\sigma^2\,P \times 10^{-6}\right)} \qquad \text{...(ii)}$$

Dividing (ii) by (i), we get

$$\frac{\lambda_1}{\lambda} = 10^6 \qquad \text{or} \qquad \lambda_1 = 10^6\,\lambda$$

$$\therefore\ \lambda_1 = 2 \times 10^{-5} \times 10^6\ \text{cm} \qquad (\because\ \lambda = 2 \times 10^{-5}\ \text{cm})$$

$$= 20\ \text{cm}$$

(ii) Mean free path at $P = \frac{p}{2}$ and T = 2T, will be

$$\lambda_2 = \frac{k.2T}{\sqrt{2}\,\pi\,\sigma^2 . \frac{P}{2}} = \frac{4kT}{\sqrt{2}\,\pi\,\sigma^2 P} \quad \ldots \text{(iii)}$$

Dividing (iii) by (i) , we get

$$\frac{\lambda_1}{\lambda} = 4 \quad \text{or} \quad \lambda_2 = 4\lambda$$

$$\therefore \lambda_2 = 4 \times 2 \times 10^{-5}$$

$$= 8 \times 10^{-5} \text{ cm}$$

Problem 32 : The mean free path of nitrogen molecules at 0°C and 1 atmosphere is 0.80×10^{-5} cm. At this temperature and pressure there are 2.7×10^{19} molecules/cm^3. Calculate the molecular diameter.

(Rohilkhand 1977)

Solution : We know,

$$\lambda = \frac{1}{\sqrt{2}\left(\pi n \sigma^2\right)}$$

$$\sigma^2 = \frac{1}{\sqrt{2}(\pi n \lambda)}$$

Given : $n = 2.7 \times 10^{19}$ mole / cm^3; $\lambda = 0.80 \times 10^{-5}$ cm

$$\therefore \sigma^2 = \frac{1}{\sqrt{2} \times 3.14 \times 2.7 \times 10^{19} \times 0.80 \times 10^{-5}}$$

$$= 10.18 \times 10^{-16} \text{cm}^2$$

or $\sigma = 3.228 \times 10^{-8}$ cm

Problem 33 : In the voltage regulator tube fixed with Argon, it is deisrable for the mean free path to be of the same order of magnitude as the spacing between the electrodes. If this distance is 0.2 cm approximately. Estimate the pressure of the gas (in mm of mercury) at room temperature 27°C. Also calculate the number of molecules per cc. of the gas is this tube. Given the diameter of Argon atom = 3.0×10^{-8} cm.

(Agra 1972s)

Solution :

We know $\lambda = \dfrac{kT}{\sqrt{2}\left(\pi\sigma^2 P\right)}$ or $P = \dfrac{kT}{\sqrt{2}\left(\pi\sigma^2 \lambda\right)}$

Given : T = 27°C = 300 K; $\lambda = 0.2$ cm; $\sigma = 3.0 \times 10^{-8}$ cm

and k = 1.38×10^{-16} ergs/K.

$$\therefore \quad P = \frac{1.38 \times 10^{-16} \times 300}{1.414 \times 3.14 \times \left(3 \times 10^{-8}\right)^2 \times 0.2} = 51.81 \text{ dynes/cm}^2$$

$$\text{or} \quad P = \frac{51.81}{13.6 \times 980} = \text{cm of Hg}$$

$$= \frac{51.81 \times 10}{13.6 \times 980} = \text{mm of Hg}$$

$$= 3.886 \times 10^{-2} \text{ mm of Hg}$$

Let the number of molecules of the gas in the tube be n, then

$$\lambda = \frac{1}{\sqrt{2}\left(\pi\sigma^2 n\right)}$$

$$\text{or} \quad n = \frac{1}{\sqrt{2}\left(\pi\sigma^2 \lambda\right)} = \frac{1}{1.414 \times 3.14 \times \left(3 \times 10^{-18}\right)^2 \times 0.2}$$

$$= 1.291 \times 10^{18} \text{ molecules per cc.}$$

Problem 34 : Calculate the mean free path of molecules in a chamber of 10^{-16} mm of Hg pressure, assuming the molecular diatemter to be 2.0 Angstron. Take the temperature of the chamber as 273 K. One gram mole of the gas occupies 22.4 litres at N.T.P. and one atmosphere is 760 mm of Hg (1 atmosphere = 1×10^6 dynes/cm^2; k = 1.4×10^{-16} ergs/K). (Agra 1975)

Solution : We know $\lambda = \dfrac{\hbar T}{\sqrt{2}\left(\pi\sigma^2 P\right)}$

Given : $T = 273K; \sigma = 2.0\ A^\circ = 2 \times 10^{-8}$ cm

$$P = 10^{-6}\text{ mm of Hg} = \frac{10^{-6}}{760}\text{ atmospheres}$$

$$= \frac{10^{-6}}{760} \times 1.0 \times 10^{6}\text{ dynes/cm}^2 = \frac{1}{760}\text{ dynes/cm}^2$$

$$k = 1.4 \times 10^{-16}\text{ ergs/K}$$

$$\therefore \quad \lambda = \frac{1.4 \times 10^{-16} \times 273}{\sqrt{2} \times 3.14 \times \left(2 \times 10^{-8}\right)^2 \times \frac{1}{760}}$$

$$= 1.63 \times 10^{4}\text{ cm}\cdot$$

Problem 35 : Calculate the frequency of sound at which the wave length of sound waves becomes equal to the mean free path of oxygen at 0°C and 1.0 atmospheric pressure. Take diameter of oxygen molecule as 3×10^{-8} cm.

Solution : We know, $\lambda = \dfrac{kT}{\sqrt{2}\left(\pi \sigma^2 n\right)}$

At N.T.P.

$$n = \frac{6 \times 10^{23}}{22400} \qquad \left[\because 22400\text{ cc. at N.T.P.} = 6 \times 10^{23}\right]$$

Given : $\sigma = 3 \times 10^{-8}$ cm

Thus
$$\lambda = \frac{1}{1.414 \times 3.14 \times \left(3 \times 10^{-8}\right)^2 \times \frac{6 \times 10^{23}}{22400}}$$

$$= \frac{22400}{1.414 \times 3.14 \times \left(3 \times 10^{-8}\right)^2 \times 6 \times 10^{23}}$$

$$= 9.37 \times 10^{-6}\text{ cm.}$$

Thus the wave length of sound waves = 9.37 × 10⁻⁶ cm

Hence frequence $= \frac{\nu}{\lambda} = \frac{332 \times 100 \text{cm/s}}{9.37 \times 10^{-6} \text{cm}}$

$$= 3.54 \times 10^{9} \text{ sec}^{-1}$$

Problem 36 : The molecular diameter of nitrogen is 3.5×10^{-8} cm. Calculate the mean free path at temperature 27°C and pressure 1 atmopshere.

(Rohtak 1977)

Solution : We know $\lambda = \frac{\hbar T}{\sqrt{2}\left(\pi \sigma^2 P\right)}$

Given : T = 27°C = 300 K; P = 1 atmosphere = 1.01×10^{-6} dynes/cm²

$$\sigma = 3.5 \times 10^{-8} \text{cm} \;\therefore\; k = 1.38 \times 10^{-16} \text{ergs/K}$$

$$\lambda = \frac{1.38 \times 10^{-16} \times 300}{1.414 \times 3.14 \times \left(3.15 \times 10^{-8}\right)^2 \times 10^{6}}$$

$$= \frac{4.14 \times 10^{-14}}{10^{6} \times 1.414 \times 3.14 \times \left(3.15 \times 10^{-8}\right)^2}$$

$$= 7.6 \times 10^{-6} \text{cm}$$

Problem 37 : Calculate the diameter of a molecule of Benzene, if n = 2.79×10^{19} molecules per cc and mean free path λ for Benzene = 2.2×10^{-6} cm.

(Agra 1986)

Solution : We know, $\lambda = \frac{1}{\sqrt{2}\left(\pi \sigma^2 n\right)}$

or $$\sigma^2 = \frac{1}{\sqrt{2}(\pi n \lambda)}$$

Given : $n = 2.79 \times 10^{19}$; $\lambda = 2.2 \times 10^{-6}$ cm

$$\therefore \sigma^2 = \frac{1}{\sqrt{2}(\pi n \lambda)}$$

$$= \frac{1}{\sqrt{2} \times 3.14 \times 2.79 \times 10^{19} \times 2.2 \times 10^{-6}}$$

$$= \frac{1}{1.414 \times 3.14 \times 2.79 \times 10^{19} \times 2.2 \times 10^{-6}}$$

$$= \frac{1}{2.725 \times 10^{14}} \text{ cm}$$

or $\sigma = 6.02 \times 10^{-8}$ cm

Problem 38 : If $\eta = 166 \times 10^{-6}$ dynes per square cm per unit velocity gradient $\overline{v} = 4.5 \times 10^4$ cm / sec; $\ell = 1.25 \times 10^{-3}$ g / cc and $n = 2.7 \times 10^{19}$ moleculesper cc for nitrogen. Calculate the mean free path, frequency of collision and the moleculare diameter of nitrogen molecules.

(Agra 1977)

Solution : For coefficient of viscosity (η), we have

$$\eta = \frac{1}{3} m\, n\, \overline{v}\, \lambda$$

$$\therefore \text{ Mean free path, } \lambda = \frac{3\eta}{m\, n\, \overline{v}} = \frac{3\eta}{\ell . \overline{v}}$$

Given : $\eta = 166 \times 10^{-6}$ dynes / sec cm per unit velocity gradient

$$m\, n = \ell = 1.25 \times 10^{-3} \text{ gm / cc}$$

$$\overline{v} = 4.5 \times 10^4 \text{ cm / sec}$$

$$\therefore \lambda = \frac{3 \times 166 \times 10^{6}}{1.25 \times 10^{-3} \times 4.5 \times 10^4} = 8.8 \times 10^{-8} \text{ cm}$$

Also Average Time between two collisions

$\lambda / \sqrt{\overline{v}}$ [mean free bath is the average distance travelled between two collisions]

$$\therefore \text{ Number of collisions / sec} = \frac{\overline{v}}{\lambda}$$

or Frequency of collisions $= \dfrac{\overline{v}}{\lambda} = \dfrac{4.5 \times 10^4}{8.8 \times 10^{-6}} = 5 \times 10^9$

since $\lambda = \dfrac{1}{\sqrt{2}\left(\pi n \sigma^2\right)}$

or $\sigma^2 = \dfrac{1}{\sqrt{2}(\pi n \lambda)}$

$$\therefore \sigma^2 = \frac{1}{\sqrt{2} \times 3.14 \times 2.7 \times 10^{19} \times 8.8 \times 10^{-6}}$$

or $\sigma = \dfrac{1}{\sqrt{1.414 \times 3.14 \times 2.7 \times 10^{19} \times 8.8 \times 10^{-6}}}$

$= 3.66 \times 10^{-8}$ cm.

Problem 39 : Determine the mean free path in nitrogen. Given that the density of nitrogen at N.T.P. = 1.2 × 10^{-3} g/cc. and that its coefficient of viscosity is 1.7 × 10^{-4} g/cm^2 per unit velocity gradient.

$\left(1 \text{ a.m.u.} = \dfrac{1}{6 \times 10^{23}} \text{gm}\right)$ (Agra 1977)

Solution : We know, $\eta = \dfrac{1}{3} m\, n\, \overline{v}\, \lambda = \dfrac{1}{3} \ell\, \overline{v}\, \lambda$ $[\because mn = \ell]$

or $\lambda = \dfrac{3\eta}{\ell\, \overline{v}}$

Also $\overline{v} = \sqrt{\dfrac{8kT}{m\pi}}$

Thus $\lambda = \dfrac{3\eta}{\ell\sqrt{\frac{8kT}{m\pi}}} = \dfrac{3\eta}{\ell}\sqrt{\dfrac{m\,n}{8kT}}$

Given : $\eta = 1.7 \times 10^{-4}$ dynes/cm^2 per unit velocity gradient

$\ell = 1.2 \times 10^{-3}$ g/cc.

$$m = 28 \text{ a m u} = \left[\frac{28}{6\times10^{23}}\right]$$

$$T = 0°C = 273K$$

$$\therefore \lambda = \frac{3\times1.7\times10^{-4}}{1.2\times10^{-3}}\sqrt{\frac{28\times3.14}{6\times10^{23}\times8\times1.38\times10^{-16}\times273}}$$

$$= 8.498\times10^{-6}\text{cm.}$$

Problem 40 : The coefficient of viscosity of helium at 273 K and 1 atmosphere is η_1, that of argon is η_2. The atomic masses of these monoatomic gases are μ_1 and μ_2 and respectivley. Calculate in terms of the above constants :

(i) The ratio $\frac{S_2}{S_1}$ where S_2 is the total scattering cross-section of Ar – Ar atom while S_1 is the total scattering cross section of the He – He atom

(ii) The ratio $\frac{lc_2}{lc_1}$ where K_2 is the thermal condutivity of argon gas while K_1 is the thermal conductivity of the heluim gas when T = 273 K

(Agra 1971s)

Solution :

We know, $$\eta = \frac{1m\bar{v}}{3\sqrt{2}\pi\sigma^2} = \frac{1}{3\sqrt{2}}\times\frac{m\,\bar{v}}{S}$$

Where S is the scattering cross-section

For Argon, $$\eta_1 = \frac{1}{3\sqrt{2}}\times\frac{m_1\bar{v}_1}{S_1}$$

and For Helium, $$\eta_2 = \frac{1}{3\sqrt{2}}\times\frac{m_2\bar{v}_2}{S_2}$$

$$\therefore \frac{\eta_1}{\eta_2} = \frac{m_1}{m_2}\times\frac{\bar{v}_1}{\bar{v}_2}.\frac{S_2}{S_1}$$

or $$\frac{S_2}{S_1} = \frac{\eta_1}{\eta_2} . \frac{m_2}{m_1} . \frac{\bar{v}_2}{\bar{v}_1} \qquad \ldots \text{(i)}$$

But $\bar{v}_1 = \sqrt{\frac{8kT}{m_1 \pi}}$ or $\bar{v}_2 = \sqrt{\frac{8kT}{m_2 \pi}}$

So that $$\frac{\bar{v}_2}{\bar{v}_1} = \sqrt{\frac{m_2}{m_1}}$$

Thus from equation (i)

$$\frac{S_2}{S_1} = \frac{\eta_1}{\eta_2} . \frac{m_2}{m_1} \sqrt{\frac{m_2}{m_1}} = \frac{\eta_1}{\eta_2} \left(\frac{m_2}{m_1} \right)^{3/2} \qquad \ldots \text{(ii)}$$

We also know $m_2 = \frac{\mu_2}{N}$ and $m_1 = \frac{\mu_1}{N}$

where N is Avagadros number

$$\therefore \frac{m_2}{m_1} = \frac{\mu_2}{\mu_1}$$

Thus from equation (ii)

$$\frac{S_2}{S_1} = \frac{\eta_1}{\eta_2} \times \left(\frac{\mu_2}{\mu_1} \right)^{3/2} \qquad \ldots \text{(iii)}$$

(ii) The thermal conductivity is given by

$$K = \frac{1}{3\sqrt{2}} . \frac{c_v}{N.\pi.\sigma^2} \times \bar{v} = \frac{1}{3\sqrt{2}} \times \frac{c_v \bar{v}}{NS}$$

Where C_v is molar or atomic specific heat and S is scattering cross-section.

As $$\bar{v} = \sqrt{\frac{8kT}{m\,\pi}} = \sqrt{\frac{8kT}{\left(\frac{\mu}{N}\right).\pi}} = \sqrt{\frac{8kTN}{\mu\pi}}$$

$$\therefore \quad K = \frac{1}{3\sqrt{2}} . \frac{c_k}{NS} \sqrt{\frac{8kTN}{\mu\pi}} \frac{1}{3\sqrt{2}} . \frac{c_v}{S} = \sqrt{\frac{8kT}{\mu N \pi}}$$

For Argon, $K_1 = \frac{1}{3\sqrt{2}} \cdot \frac{c_v}{S_2} \sqrt{\frac{8\,kT}{\mu_1 N\,\pi}}$

For Helium, $K_2 = \frac{1}{3\sqrt{2}} \cdot \frac{c_v}{S_2} \sqrt{\frac{8\,kT}{\mu_2\,\pi\,N}}$

$$\therefore \frac{K_2}{K_1} = \frac{S_1}{S_2}\sqrt{\frac{\mu_1}{\mu_2}} \qquad \ldots \text{(iv)}$$

From eqaution (iii)

$$\frac{S_1}{S_2} = \frac{\eta_2}{\eta_1} \times \left(\frac{\mu_1}{\mu_2}\right)^{3/2}$$

$$\therefore \quad \frac{K_2}{K_1} = \frac{\eta_2}{\eta_1} \times \left(\frac{\mu_1}{\mu_2}\right)^{3/2} \sqrt{\frac{\mu_1}{\mu_2}}$$

$$\text{or} \quad \frac{K_2}{K_1} = \frac{\eta_2}{\eta_1} \times \left(\frac{\mu_1}{\mu_2}\right)^{2} \qquad \ldots \text{(v)}$$

14

Behaviour of Real Gases

Problem 1 : The constants 'a' and 'b' in Vander Waals' equation of state for CO_2 are 0.364 N $m^4/mole^2$ and 42.8 cm^2/mole respectively. Calculate the pressure of this gas to 0°C, when the volume is 0.560 litre/mole. Assume that Vander Waals' equation holds good. Compare this the pressure with pressure given by ideal gas equation. What is the internal pressure ?

Solution :

$$\left(P+\frac{a}{v^2}\right)(v-b)=R_0T$$

or $$\left[P+\frac{0.364}{\left(0.560\times10^{-3}\,m^3mol^{-1}\right)^2}\right]$$

$$\left[0.560\times10^{-3}\,m^3mol^{-1}-42.8\times10^{-6}\,m^3mol^{-1}\right]$$

$$=8.314\times Jk^{-1}mol^{-1}\times273.15k$$

or $$=\left(P+1.161\times10^6\,N\,m^{-2}\right)=4.391\times10^6\,N\,m^{-2}$$

$$\therefore\quad P=3.23\times100^6\,N\,m^{-2}$$

$$\text{or }=\frac{3.23\times100^6}{1.013\times10^5}\text{ atmo}=31.9\text{ atmospheres}$$

However for a prefect gas, we have Pv = RT

$$\text{or} \qquad P = \frac{8.314 \text{Jk}^{-1}\text{mol}^{-1} \times 273.15\text{k}}{0.560 \times 10^{-3} \text{m}^3 \text{mol}^{-1}}$$

$$= 4.05 \times 10^6 \text{ N m}^{-2}$$

= 40 atmospheres.

$$\text{Internal Pressure} = \frac{a}{v^2}$$

$$= 1.161 \times 10^6 \text{ N m}^{-2}$$

= 11.5 atmo

Problem 2 : Calculate the value of critical temperature for carbon dioxide.

Given a = 0.00874 atmo; b = 0.0023 cc.

Solution : Assuming P = 1 atmosphere, v = 1cc and T = 0°C we can find out the values of R by using the eqn.

$$\left(P + \frac{a}{v^2}\right)(v - b) = RT$$

$$\left(P + \frac{0.00874}{1^2}\right)(1 - 0.003) = R \times 273$$

$$\text{or} \qquad R = \frac{1.00874 \times 0.9977}{273}$$

$$\text{Hence} \qquad T_c = \frac{8a}{27Rb} = \frac{8 \times 0.00874 \times 273}{27 \times 1.00874 \times 0.9977 \times 0.0023}$$

$$= 305\text{k} \quad \text{or} \quad 32.5°\text{C}$$

Problem 3 : For one mole of hydrogen, the Vander Waals' constant are

$a = 0.245$ litre^2 – atom–mol^{-2}

$b = 2.67 \times 10^{-2}$ litre mol^{-1}

Calculate its temperature of inversion. Given that R = 2 cal mol^{-1} k^{-1} and 1 atmosphere pressure = 10^6 dynes/cm^2

[Rajasthan 1979]

Solution : The temperature of inversion T_i is given by

$$T_i = \frac{2a}{R-b}$$

Substituting the given values of a, b and k,

we get $$T_i = \frac{2 \times 0.245 \times 10^{12}}{2 \times 4.2 \times 10^7 \times 26.7} = 220k$$

$[\because\ a = 0.245 \text{ litre}^2 \text{ atom mol}^{-2}$

$= 0.245 \times 10^6 \times 10^6 \text{ cm}^4 \text{ dynes mol}^{-2}$

$b = 2.67 \times 10^{-2} \text{ litre mol}^{-1}$

$= 26.7 \text{ cm}^3, \text{ mol}^{-1}$

$R = 2 \text{ Cal mol}^{-1} = 2 \times 4.2 \times 10^7 \text{ ergs/mole k}]$

Problem 4 : Calculate the critical temperature and Boyle temperature of CO_2, for which the Vander Waals constants are given to be

$$a = 0.0072 \text{ and } b = 0.002$$

The unit of pressure is atmosphere and the unit of volume is that of a gm mol of the gas at N.T.P.

Solution : From Vander Waals equation, we have

$$\left(P + \frac{a}{v^2}\right)(v - b) = RT$$

At N.T.P., P = 1 and v = 1 (given)

$$\therefore \quad \left(1 + \frac{0.0072}{1^2}\right)(1 - 0.002) = R \times 273$$

Thus $$R = \frac{1.0072 \times 0.998}{273}$$

Since critical temp. $$T_c = \frac{8a}{27R.b}$$

$$T_c = \frac{8 \times 0.0072}{27 \times \left(\frac{1.0072 \times 0.998}{273}\right) \times 0.002}$$

$= 289.6k$ or $= 16.6°C$

The Boyle Temperature = $T_B = \frac{6.75}{2} T_c$

$$\left[\begin{array}{r} \because\ 2T_B = 6.75T_c \\ T_i = 2T_B \end{array}\right]$$

$$= \frac{6.75}{2} \times 289.6$$

$$= 977.4\text{k} \quad \text{or} \quad = 664.4°\text{C}$$

Problem 5 : Show that for a gas obeying Vander Waals equation

$$\frac{RT_c}{P_c v_c} = \frac{8}{3}$$

Where the symbols have their usual meanings.

[Meerut 1987]

Solution : We have

$$v_c = 3b$$

$$T_c = \frac{8a}{27R.b}$$

$$P_c = \frac{8}{27b^2}$$

$$\therefore \quad \frac{RT_c}{P_c v_c} = \frac{R \times \frac{8a}{27R.b}}{\left(\frac{8}{27b^2}\right)(3b)} = \frac{8}{3}.$$

Problem 6 : Calculate the critical temperature of a gas for which the Vander Waals constants are

$$a = 0.00874,\ b = 0.0023$$

and the gas constant is given by

$$273R = 1.00646$$

[Meerut 1982]

Solution : $T_c = \frac{8a}{27R.b}$

$$= \frac{8 \times 0.00874}{27 \times \left(\frac{1.00646}{273}\right) \times 0.0023}$$

$$= 305.5°\text{C}$$

15

Change of State

Problem 1 : A closed vessel contains a mixture of air and water vapour in contact with excess of water. The pressure in the vessle at 27°C and 60°C are respectively 77.7 cm and 98.1 cm of mercury. If the vapour pressure of water at 27°C in 2.7 cm of mercury, what is the vapour pressure at 60°C?

Solution : The pressure of air at 27°C = 77.7 – 2.7 = 75.0 cm

The pressure of air at 60°C = 98.1 – p

where p is the saturated vapour pressure of water at 60°C

At constant volume, P and T (For constant mass)

For constant mass of air in mixture

$$\frac{98.1-P}{75.0}=\frac{273+60}{273+27}$$

or $$98.1-P=\frac{333}{300}\times 75.0$$

or $$P=14.85 \text{ cm.}$$

Problem 2 : The saturated vapour pressure of water at 11.25°C is 1 cm of mercury, the density of dry air at 76 cm pressure and 25°C is 0.001184 g/cm^3, and the density of water vapour relative to that of air at the same pressure and temperature is 0.624. Find the weight of a litre of moist air at 25°C, if the barometer stands at 76 cm and dew point is 11.25°C.

Solution : Since the dew point is 11.25°C, the saturated vapour pressure of the water = 1 cm of mercury. Hence the pressure of air at 25°C = (76 – 1) = 75 cm mass of 1000 cc of air at 76 cm pressure and 25°C temperature = 1.184g (given 0.001184 × 1000 = 1.184 g).

At 75 cm pressure and 25°C, the volume of this air $= 1000 \times \frac{76}{75}$ cc.

[From Boyles Law]

$$\therefore \text{ mass of 1 litre of air at 75 cm and } 25°\text{c} = \frac{1.184 \times 1000}{1000 \times \frac{76}{75}}$$

$$= 1.184 \times \frac{75}{76}$$

$$= 1.169\text{g}$$

To find the mass of 1 litre of water vapour at 1 cm and 25°C

We know

Pressure of water–vapour in moist air = 1 cm

Mass of 1 litre of air at 1 cm and 25°C $= \frac{1}{75} \times 1.169\text{g}$

Density of water vapour = 0.624 of that of air (given)

∴ mass of 1 litre of water vapour at 1 cm pressure and 25°C

$$= 0.624 \times \frac{1}{75} \times 1.69$$

$$= 0.0097\text{g}$$

∴ mass of moist air = (1.169 + 0.0097) g

= 1.179 g.

Problem 3 : Calculate what fraction of the mass of water vapour would condense if the temperature fell from 20°C to 5°C, and if originally at 20°C, the humidity was 60%.

(Saturation vapour pressure of water vapour at 20°C = 17.5 mm and saturation vapour pressure of water vapour at 5°C = 6.5 mm)

Solution : Let the mass of water vapour in 1 *l* of air at 20°C = m and the mass of water vapour required to saturate 1 *l* of air at 20°C= M.

Then relative humidity $= \frac{m}{M} \times 100 = 60\%$ (given)

$$\text{or} \qquad m = 0.6M = \frac{3}{5}M.$$

If m_1 is the mass of the water vapour required to saturate 1 *l* of air at 5°C, then

$$\frac{m_1}{M} = \frac{\text{saturation vapour pressure at } 5°\text{C}}{\text{saturation vapour pressure at } 20°\text{C}}$$

$$\text{or} \qquad \frac{m_1}{M} = \frac{6.5}{17.5} \qquad \therefore m_1 = \frac{13}{35}M$$

Thus the mass of water–vapour in 1 litre which condenses when the temperautre changes from

$$20°\text{C to } 5°\text{C} = m - m_1 = \frac{3}{5}M - \frac{13}{35}M.$$

$$= \frac{8}{35}M$$

∴ Fraction condensed

$$= \frac{\frac{8}{35}M}{\frac{3}{5}M} = \frac{8}{21}.$$

Problem 4 : Calculate the temperature for triple point of water in the absence of air. Given that the vapour pressure of air at 0°C and 1°C are 4.60 mm and 4.94 mm respectively and lowering of melting point per atmosphere is 0.0072°C.

Solution :

Saturated vapour pressure at 0°C = 4.60 mm

Saturated vapour pressure at 1°C = 4.94 mm

∴ Increase in vapour pressure per °C = 0.34 mm

Hence the saturated vapour pressure at t°C = (4.60 + 0.34 t)

= p mm

or $p = 4.6 + 0.34\,t$

or $$t = \frac{p - 4.6}{0.34}$$

Substituting the value of t in the relation

$$t = 0.0072 - \frac{0.0072}{760}p$$ [For triple point of water]

We have

$$\frac{p-4.6}{0.34} = \left(0.0072 - \frac{0.0072}{760}p\right)\text{mm}$$

or $p = 4.6024$ mm

Hence $(4.60 + 0.34\,t) = 4.6024$

or $t = 0.007156$°C

16

Specific Heats of Solids and Liquids

Problem 1 : 200 g of water and an equal volume of another liquid of mass 250 grams are placed in turn in the same calorimeter of 100 gm and specific heat 0.1. The liquids which are constantly stirred, are formed to cool from 60°C to 40°C in 180 sec and in 140 sec respectively. Find the specific heat of the liquid.

Solution : The average heat lost by the liquid and the calorimeter while the temperature falls from 60°C to 40°C is equal to the average heat lost by the water and calorimeter from 60°C to 40°C.

Since the excess temperature over the surroundings, and the nature and area of the cooling surfaces, are the same in both cases.

(Please note that Newton's law of cooling in not being used).

The average rate of loss of heat of the liquid and the calorimeter in cal per sec

$$= \frac{(200s + 100 \times 0.1)\ (60 - 40)}{140}$$

and the average rate of loss of heat of water and the calorimeter in cal per sec =

$$= \frac{(200 + 100 \times 0.1)\ (60 - 40)}{180}$$

$$\therefore \quad \frac{(250s + 10) \times 20}{140} = \frac{210 \times 20}{180}$$

$$\therefore \quad s = 0.61 \text{cal}/\text{gm}/°\text{C}$$

Problem 2 : A copper calorimeter of mass 100 gm containing 150 cc of a liquid of specific heat 0.6 and specific gravity 1.2, is found to cool at the rate of 2°C per minute when its temperature is 50° above that of its surroundings. If the liquid is emitted out, and 150 cc of a liquid of specific heat 0.4 and specific gravity 0.9 are substituted, what will be the rate of cooling when the temperature is 40° above that of surroundings? (specific heat of copper = 0.01)

Solution :

The mass of the first liquid = 150×1.2 = 180 g

∴ heat lost/min at 50°C above surroundings

$$= (180 \times 0.6 + 100 \times 0.1) \times 2 \text{ cal.}$$

The mass of the second liquid = 150 × 0.9=135 g

∴ heat lost/min at 40°C above surroundings

$$= (135 \times 0.4 + 100 \times 0.1) \times x \text{ cal.}$$

Where x is the rate of cooling. Assuming *Newton's Law of Cooling*, the rate of loss of heat is proportional to the excess temperature over the surroundings.

$$\therefore \quad \frac{(135 \times 0.4 + 100 \times 0.1) \times x}{(180 \times 0.6 + 100 \times 0.1) \times 2} = \frac{40}{50}$$

$$\text{or} \quad 64x = \frac{40 \times 118 \times 2}{50} = 2.95°\text{C per min.}$$

Problem 3 : In an experiment with callendar and Barne's continuous flow method the potential difference across the wire was 2.5 volts and the current flowing was 2 amperes, the rise in temperature of the liquid is 5°C when the rate of flow is 32 g/min. The rate of flow was then increased to 50 g/min, the potential difference to 3 volts and current to 2.5 amp, the temperture difference being the same as before. Calculate the specific heat of the liquid. Given J = 4.2 Jouls/cal,

Solution : Using the formula

$$\text{Specific heat } s = \frac{(E'I' - EI)t}{J(m' - m)(Q_2 - Q_1)}$$

and substituting the given values, we get

$$s = \frac{(3.0 \times 2.5 - 2.5 \times 2.0) \times 60}{(50 - 32) \times 5 \times 4.2}$$

$$= 0.4$$

Problem 4 : A copper calorimeter of mass 120 gm contains 150 g of a liquid at 26°C. 25g of ice at 0°C are added, and the minimum temperature of the mixture is recorded. At this temperature, vapour from the same boiling liquid is introduced into the mixture until the final temperature of the calorimeter and new content is 30°C. What mass of vapour is condensed?

(Specific heat of copper = 0.6 cal/gm/°C; b.p of liquid = 64.7°C; Latent heat of evaporation of liquid = 267 cal/gm; Latent heat of fusion of ice = 80 cal/gm)

Solution :

Let the minimum temperature of the mixture is t°C.

Since

Heat lost by liquid and calorimeter = Heat gained by ice and water formed.

$$(150 \times 0.6 + 120 \times 0.1)(26 - t) = 25 \times 80 + 25 \times (t - o)$$

or $t = 5.1°C$

Let the mass of vapour condenced = m

Since

Heat given out by condensed vapour and liquid formed = Heat gained by liquid, water and calorimeter

$\therefore$ $267\ m + m \times 0.6 \times (64.7–30.0) = (150 \times 0.6 + 25 + 120 \times 0.1)$ $(30–5.1)$

$\therefore$ $287.82\ m = 3162.3$

or $m = 10.99$ g

Problem 5 : When a quantity of liquid bismuth at its melting point is transferred to a calorimeter containing oil, the temperature of oil rises from 12.5°C to 27.6°C. When the experiment is repeated with all circumstances the same except bismuth is solid, the temperature of the oil rises to 18.1°C. The specific heat of bismuith is 0.032 cal per gram per °C. What is the latent heat of fusion ?

Solution : Let W be the water equivalent of the calorimeter in grams, m_1, m_2 the respective masses in grams of bismuth and oil, and s the specific heat of the oil. When the liquid bismuth is transferred, it solidifies inside the oil, because the temperature is lower than the melting point. Since the heat given out by the bismuth on solidification plus that given out by the solid bismuth when its temperature falls from 271°C to 27.6°C is equal to the heat absorbed by the calorimeter and contents.

$$\therefore\ m_1L + m_1 \times 0.032 \times (271 - 27.6) = (m_2s + W)(27.6 - 12.5)$$

$$\therefore\ m_2(L + 0.032 \times 243.4) = (m_2s + W) \times 15. \qquad ...(i)$$

When the solid bismuth at 271°C, is transferred, no change of state occurs.

Since the heat given out by solid

= heat absorbed by the calorimeter and contents.

$$\therefore\ m_1 \times 0.032 \times (271 - 18.1) = (m_2s + W)(18.1 - 12.5)$$

$$\therefore\ m_1 \times 0.032 \times 252.9 = (m_2s + W) \times 5.6 \qquad ...(ii)$$

Dividing (i) by (ii) to eliminate all unknown quantities expcet L,

$$\therefore\ \frac{L + 0.032 \times 243.4}{0.032 \times 252.9} = \frac{15.1}{5.6}$$

$$\because\ 0.032 \times 243.4 \times 5.6L = 0.032 \times 252.9 \times 15.1$$

$$\therefore\ L = 14.0 \text{ cal per gram}$$

Problem 6 : The capillary tube of a Bunsen ice calorimeter has an internal diameter of 0.4 mm. When a piece of metal of mass 0.5 g heated to 100°C is dropped into the calorimeter, the mercury meuiscus moves 4 cm. What is the specific heat of the metal? (sp. gravity of water at 0°C = 1.00, sp. gravity of ice at 0°C = 0.917; latent heat of fusion of ice = 80 cal/gm)

Solution :

1 g of water has a volume = 1.00 cc

1 g of ice has a volume = 1.091 cc $\left(\because v = \frac{1}{0.917}\right)$

From this we find that a contraction in volume when 1 g ice melts

= 0.091 cc.

Since contraction $= \pi r^2 l$

$= \pi \times (0.02)^2 \times 4 = 0.0016\, \pi$ cc.

$\therefore$ the mass of ice melted $= \frac{0.0016\pi}{0.091}$ gm

$\therefore$ heat given out by metal $= \frac{0.0016\pi \times 80}{0.091}$ cal

$\therefore$ specific heat of metal $= \frac{Q}{m.t}$

$$= \frac{0.0016\pi \times 80}{0.091 \times 0.5 \times 100}$$

$$= 0.088 \text{cal} / \text{gm} / ^\circ\text{C}$$

Problem 7 : Water in a vacuum flask is boiled steadily by passing an electric current through a coil of wire immersed in the water. When the potential difference across the coil is 5.25 V and the current through it is 2.58 amp, 6.85g of water evaporates in 20 min. When the potential difference and the current are maintained at 3.20V and 1.57 amp respectively, 2.83 g of water evaporates in 20 min, all the other conditions being the same. Calculate the latent heat of vaporisation of water in Jolues per gram.

Solution : If I is the current in amperes in a coil of wire and V is the potential difference in volts across the coil, the heat produced per second in the coil is **IV Jolues.**

This heat is expended in evaporating the water and in making up the heat losses. Thus

$$2.58 \times 5.25 = \frac{6.85L}{20 \times 60} + h \qquad \ldots \text{(i)}$$

Where L = Latent heat of vaporisation in Joules/gm

h = heat lost/sec in Joules

In the second case

$$1.57 \times 3.20 = \frac{2.382}{20 \times 60} + h \qquad \ldots \text{(ii)}$$

As the conditions of cooling are the same.

Substracting (ii) from (i), we get

$$(2.58 \times 5.25 - 1.57 \times 3.20) = \frac{6.85 - 2.38}{20 \times 60} L$$

$$\text{or} \qquad L = \frac{1200(2.58 \times 5.25 - 1.57 \times 3.20)}{4.47}$$

$$= 2290 \text{ Joules / g}$$

17

Specific Heat of Gases

Problem 1 : A gas initially it S.T.P. and a volume of 1000 cc expands to double its volume without change of pressure. Calculate the quantity of heat that must be supplied.

Given : Density of mercury = 13.6 g/cc.

Sp. heat of gas at constant volume, Cv = 0.168 cal/gm

desity of gas at S.T.P. = 0.00129 g/cc

$J = 4.18 \times 10^7$ ergs/cal.

$g = 981$ cm/sec^2

Solution :

The heat required is given by C_p m.t., where

m = mass of the gas

C_P = specific heat of gas at constant pressure

and t = rise in temperature.

At constant pressure, for a gas, we have; $v \alpha T$

$$\frac{273+x}{273}=\frac{2}{1}$$

where x is the final temperature of the gas after expansion

$$\therefore \quad 273+x=2\times 273$$

$$\text{or} \quad x=273°C$$

$$\therefore \quad t = (273-0)°C$$

$$=273°C$$

Also $C_p = C_v + R$ (For one mole)

For 1g we have $R = \frac{76 \times 13.6 \times 981 \text{ ergs/g/°C}}{273 \times 0.00129}$ $\left[\because R = \frac{Pv}{T}\right]$

Thus $C_p = c_v + \frac{R}{J}$

$$= 0.168 + \frac{76 \times 13.6 \times 981}{273 \times 0.00129 \times 4.18 \times 10^7}$$

$\therefore C_p = 0.237$ cal/gm/°C

Heat supplied = 1.29×0.237×273=83.5 cal (Heat C_p m.t.)

Problem 2 : 100 cc of air at a pressure of 75 cm of mercury is compressed adiabatically till its volume becomes 25 cc. What is the new pressure of the air ? [Given : g = 1.4]

Solution : For adiabatic process pv^γ = constant.

$$\therefore \; p \times (25)^{1.4} = 75 \times (100)^{1.4}$$

or $$p = \frac{75 \times (100)^{\frac{1}{4}}}{(25)^{\frac{1}{4}}} -$$

or $$\log p = \log 75 + 1.4 \log 100 - -\frac{1}{4} \log 25$$

$$= 2.7180$$

p = 522.4 cm of mercury

Problem 3 : The volume of a given mass of air is 80 cc at a pressure of 60 cm of mercury. If the gas expands adiabatically to a final pressure of 45 cm of mercury. What is the new volume of air ?

[Given $\gamma = 1.4$]

Solution : For an adiabatic process pv^r = constant

$$\therefore \; 45 \times (v)^{1.4} = 60 \times (80)^{1.4}$$

$$\therefore \; (v)^{1.4} = \frac{60 \times (80)^{1.4}}{45}$$

or $1.4 \log v = \log 60 + 1.4 \log 80 - \log 45$

or $$\log v = \frac{\log 60 + 1.4 \log 80 - \log 45}{1.4}$$

$$= 1.9924$$

or $$v = 98.3 \text{ cc}$$

Problem 4 : A certain mass of air at 40°C and a pressure of 70 cm of mercury is compressed adiabatically initial its volume is halved. What is the new pressure [$\gamma = 1.4$]

Solution : Let the original volume of air = v cc

Then its new temperature (T) is given by use of relation

$Tv^{\gamma-1}$ = constant (For adiabatic process)

i.e. $$T\left(\frac{v}{2}\right)^{\gamma-1} = 313 \times v^{\gamma-1} \qquad [\because 40°C = 313 \text{ k}]$$

$$T = \frac{313 \times v^{\gamma-1}}{\left(\frac{v}{2}\right)^{\gamma-1}} = 313 \times 2^{\gamma-1}$$

$$= 313 \times 2^{1.4-1} \qquad [\ \gamma = 1.4\ ]$$

$$= 313 \times 2^{0.4} = 413\text{k. or } 140°\text{C}$$

Now the new pressure of the air can be calculated with the help of the relation.

PV^{γ} = constant (For adiabatic process)

$$\therefore \quad P\left(\frac{v}{2}\right)^{1.4} = 70 \times (v)^{1.4}$$

or $$P = 70 = \frac{(v)^{1.4}}{\left(\frac{v}{2}\right)^{1.4}} = 70 \times 2^{1.4} = 184 \text{ cm of mercury}$$

Problem 5 : 22.4 *l* of air at 15°C and 76 cm of mercury pressure, weighing 27.3 g expands adiabatically until the volume has increased by 50 per cent.

(a) What will be final pressure and temperature ?

(b) How much work is done during expansion ?

(Given $C_v = 0.17$; $C_p = 0.24$ cal gm^{-1} $°C^{-1}$).

Solution :

(a) Final volume $= \left(22.4 + \frac{22.4 \times 50}{100}\right)$ litres

$= 33.6$ litres

Since $\gamma = \frac{C_p}{C_v}$

$\therefore \gamma = \frac{0.24}{0.17} = 1.4$ (approx)

Now for an adiabatic process Pv^γ = constant

$$\therefore \quad P \times (33.6)^\gamma = 76 \times (22.4)^\gamma$$

$$\text{or} \quad P \times (33.6)^{1.4} = 76 \times (22.4)^{1.4} \qquad [\ \gamma = 1.4]$$

$$\text{or} \quad P = 76 \times \frac{(22.4)^{1.4}}{(33.6)^{1.4}}$$

$$\text{or} \quad \log P = \log 76 + 1.4 \log 22.4 - 1.4 \log 33.6$$

$$= 1.6343$$

$$\therefore P = 43.1 \text{ cm of mercury}$$

Again for an adiabatic process, $Tv^{\gamma-1}$ = constant

$$\therefore \quad T \times (33.6)^{1.4-1} = 288 \times (22.4)^{1.4-1}$$

where T is the final temperature

$$\therefore \quad T = 288 \times \left(\frac{22.4}{33.6}\right)^{0.4} = 245° k = -28°C$$

(b) For an adiabatic process (q = 0) $\therefore\ \Delta E = W$

Now change in internal energy (ΔE) = mass $\times\ C_v \times t$

$$= 27.3 \times 0.17 \times [15 - (-28)]$$

$$= 27.3 \times 0.17 \times 43$$

$$= 199.6 \text{ Calories}$$

Problem 6 : A certain volume of dry air at N.T.P. is expanded to 3 times its volume under adiabatic conditions. Find the final pressure and temperature.

(Given : $\gamma = 1.4$)

Solution : For an adiabatic process

$$P_V{}^{\gamma} = \text{constant}$$

$$\therefore\ 76 \times v_1{}^{1.4} = P_2(3v_1)^{1.4}$$

or $$P_2 = \frac{76}{(3)^{1.4}} = 16.32\text{cm}$$

Again for an adiabatic process $Tv^{\gamma-1}$ = constant

$$T_1 v_1^{\gamma-1} = T_2 v_2^{\gamma-1}$$

or $$273 \times (v_1)^{1.4-1} = T_2(3v_1)^{1.4-1}$$

or $$T_2 = \frac{273}{(3)^{0.4}} = 176\text{k} \quad \text{or} \quad -97°\text{C}$$

Problem 7 : Calculate the final temperature when a given mass of a gas at 0°C is suddenly compressed to a pressure 20 times the initial pressure (Given : $\gamma = 1.42$)

Solution : For an adiabatic process $T^{\gamma}P^{1-\gamma}$ = constant

$$\therefore\quad T_1^{\gamma} P_1^{\gamma-1} = T_2^{\gamma} P_2^{1-\gamma}$$

or $$\left(\frac{T_2}{T_1}\right)^{\gamma} = \left(\frac{P_1}{P_2}\right)^{1-\gamma}$$

$$\text{or} \qquad T_2 = T_1\left(\frac{P_1}{P_2}\right)^{\frac{1-\gamma}{\gamma}}$$

$$= 273\left(\frac{1}{20}\right)^{\frac{1-1.42}{1.42}}$$

$$= 273\left(\frac{1}{20}\right)^{\frac{0.42}{1.42}}$$

$$\text{or} \qquad \log T_2 = \log 273 + \frac{0.42}{1.42}\log 20$$

$$= 2.8210$$

$$\therefore \quad T_2 = 662.2\text{k} \quad \text{or} \quad 399.2°\text{C}$$

Problem 8 : (i) Calculate Einstein's frequency for copper for which $\theta_E = 230$ k. Given h = 6.6 × 10^{-34} Joule–sec; Boltzmann constant k = 1.37 × 10^{-23} Joule/k.

(ii) Use ressult (i) to show heat the classical theory result $C_v = 3$ R should be valid for copper if T > 230°C

[Agra 1976]

Solution : We know, Einstein's frequencey, V_E is

$$V_E = \frac{k\theta_E}{h}$$

Given : k = 1.37 × 10^{-23} Joule/k

h = 6.6 × 10^{-34} Joule/sec

$\theta_E = 230$ k

$$V_E = \frac{230 \times 1.37 \times 10^{-23}}{6.6 \times 10^{-34}}\,\text{sec}^{-1}$$

$$= 4.8 \times 10^{12}\,\text{sec}^{-1}$$

(ii) $$C_v = 3R \text{ if } T \simeq \frac{h\nu}{h}$$

$$\frac{h\nu}{h} = \frac{6.6 \times 10^{-34} \times 4.8 \times 10^{12}}{1.37 \times 10^{-23}} = 230\text{k}.$$

Problem 9 : Calculate Einstein's frequency for a case for which $Q_E =$ 236 k.

Given : k = 1.4 × 10^{-23} Joule/k, h = 6.6 × 10^{-34} Joule – sec

[Agra 1985]

Solution : We know $\frac{\theta_E . h}{h} = V_E$

$$\therefore \quad V_E = \frac{236 \times 1.4 \times 10^{-23}}{6.6 \times 10^{-34}}$$

$$= 5.0 \times 10^{12} \text{ sec}^{-1}$$

Problem 10 : For copper V = 2.49 × 10^{12} cycles/sec (Hertz). Calculate the characteristic temperature θ_E

[Rajasthan 1981s]

Solution : We know, $\theta_E = \frac{h\nu}{k}$

$$\theta_E = \frac{6.6 \times 10^{-34} \times 2.49 \times 10^{12}}{1.38 \times 10^{-23}} = 119.1 \text{ k.}$$

Problem 11 : For copper the value C_v is 23.8 Joule/mole k while for diamond the value of C_v is only 6.1 Joule/mole k (both values are at T = 300 k). How do you account for this difference?

Given : h = 6.6 × 10^{-34} Joule/sec, k = 1.38 × 10^{-23} Joule/k.

[Agra 1970s]

Solution : According to Einstein's theory, the specific heat of solid,

$$C_v = 3R\left[\frac{e^{\frac{\theta}{T}}}{\left(e^{\frac{\theta}{T}} - 1\right)^2}\left(\frac{\theta}{T}\right)^2\right]$$

where $Q = \frac{h\nu}{h}$ is known as Einstein's temperature.

For same temperature the difference in values of C_v for copper and diamond is due to the difference in their V values.

Problem 12 : For copper atomic weight = 63.5, density 8.95 g cm^{-3} and compressibility $\beta = 7.3 \times 10^{-13}$ $cm^2/dyne^{-1}$. Using these data and the universal constant N, k and h, deduce the following :

(i) mass m of copper atom

(ii) the interatomic distance a assuming simple cubic lattice.

(iii) frequency of vibration $= \dfrac{1}{2\pi}\sqrt{\dfrac{3a}{\beta m}}$

(iv) use result (iii) to show that classical theory result $C_v = 3R$ should be valid for copper if $T \leq 230$ k

[Agra 1972]

Solution :

(i) Mass of copper atom, $m = \dfrac{\text{Atomic weight}}{\text{Avagadros number}}$

$$= \frac{63.5}{6.02 \times 10^{23}} = 1.05 \times 10^{-22}\text{ g}$$

(ii) Interatomic distances, $a = \left(\dfrac{nA}{N\ell}\right)^{\frac{1}{3}}$

Here n = Lattice points per unit cell = 1 for simple cubic lattice

A = Atomic weight = 63.5

N = Avagadros number = 6.02×10^{23}

l = density of copper = 8.95 g cm^{-3}

$$a = \left(\frac{1 \times 63.5}{6.02 \times 10^{23} \times 8.95}\right)^{\frac{1}{2}}$$

$$= 2.28 \times 10^{-8}\text{ cm} = 2.28\text{A}°.$$

(iii) $= \dfrac{1}{2\pi}\sqrt{\dfrac{3a}{\beta m}}$

$$= \frac{1}{2 \times 3.14}\sqrt{\left(\frac{3 \times 2.281 \times 10^{-8}}{7.3 \times 10^{-13} \times 1.05 \times 10^{-22}}\right)}$$

$$= 4.75 \times 10^{12}\text{ cycles/ sec}$$

(iv) We know that C_v approaches the value 3R

$$\text{if} \quad \frac{h\nu}{k} << 1 \text{ or } \frac{h\nu}{k} < T \text{ or } T > \frac{h\nu}{k}$$

using the given values we find

$$\frac{h\nu}{k} = \frac{6.02 \times 10^{-34} \times 4.75 \times 10^{12}}{1.38 \times 10^{-23}} = 230\text{k}$$

Thus for copper the result $C_v = 3R$ is valid

Problem 13 : The specific heat of a certain subtance at very low temperature is found to depend on the absolute temperature T according to the relation

$C = AT^3$, where A is a constant

(i) If units are caloric, gram and kelvin degree. What units must the constant A have?

(ii) How much heat must be added to raise the temperature of a mass *m* of the material from T_1 to T_2?

[Agra 1972s]

Solution :

(i) Since $C = AT^3$

$$\therefore A = \frac{C}{T^3}$$

Thus units of $A = \dfrac{\text{Cal Per gram per kelvin}}{(\text{kelvin})^4}$

$= \text{cal per gram per } (k)^4$

$= \text{cal gm}^{-1} \text{ k}^{-4}$

(ii) Heat need to raise the temperature of mass *m* through a temperature dT is mcdT

$\therefore$ Heat required to raise the temperature from T_1 to T_2 is given by

$$\int_{T_1}^{T_2} \text{mc dt} = \int_{T_1}^{T_2} \text{m } AT^3 \text{ dT}$$

$$= \text{m A}\left[\frac{T^4}{4}\right]_{T_1}^{T_2} = \frac{1}{4} \text{ m A} \left(T_2^4 - T_1^4\right)$$

Problem 14 : Calculate in terms of kT the vibrational contribution of energy of 1 g mole of hydrogen at a temperature at which $\frac{hV}{kT} = 2$. Given : Vibrational constant $\bar{v} = 4400\ cm^{-1}$. Calculate this temperature. Given : Avagadros Number, N = 6.0×10^{23} g–mole and

$e = 2.7183, h = 6.6 \times 10^{-27}$ ergs –sec; $k = 1.38 \times 10^{-16}$ ergs/k.

[Agra 1970]

Solution : The vibrational contribution to energy of 1 molecule of hydrogen is given by

$$\overline{E} = \frac{h\nu}{\ell^{\frac{h\nu}{kT}} - 1} = \frac{h\nu / kT}{\ell^{h\nu/kT} - 1} . kT$$

∴ vibrational contribution to energy of 1 gm mole of hydrogen is

$$E_v = N . \frac{\frac{h\nu}{kT}}{e^{\frac{h\nu}{kT}} - 1} . kT, \text{ where N = Avagadros Number}$$

$$\text{Given } = \frac{h\nu}{kT} = 2, N = 6 \times 10^{23}$$

$$\therefore \quad E = 6 \times 10^{23} \times \frac{2}{e^2 - 1} . kT$$

$$= \frac{6 \times 10^{23} \times 2}{(2.7183)^2 - 1} . kT$$

$$= 1.874\ kT$$

$$\text{Since} \quad \frac{h\nu}{kT} = 2$$

$$T = \frac{h\nu}{2k} = \frac{h c \bar{\nu}}{2k}$$

$$= \frac{6.6 \times 10^{-27} \times 3 \times 10^{10} \times 4400}{2 \times 1.38 \times 10^{-16}} = 3179\text{k}$$

18

Thermal Radiation

Problem 1 : A metal (tungsten) sphere has a diameter of 4.0 cm and an emissivity of 0.35. Find out the energy lost by radiation in one second when the temperature of the ball is 2927°C.

Solution : Energy radiated per second

$$E = e\,\sigma\, AT^4$$

where e = relative emittance (or emissivity)

σ = Stefan = Boltymann Constant

A = Surface area of sphere.

$$E = 0.35 \times 5.670 \times 10^{-8}\ Wm^{-2}k^{-4} \times 4\pi\left(\frac{2}{100}m\right)^2 \times (3200k)^4$$

$$E = 0.35 \times 5.670 \times 4\pi \times 4 \times (32)^4 \times 10^{-4}\ W$$

$$= 10.5kW$$

Problem 2 : A blackened metal container of negligible heat capacity is filled with water. The container in cubical with each side 5 cm long. It is placed in an evacuated enclosure whose walls are kept at 27°C. In what time will the water temperature change from 30°C to 29°C?

Solution : As the container is in vacuum, the energy is lost by radiation only.

$$\text{Thus, } E = \left(4\ e\,\sigma T_1^2\right)\Delta T$$

From this we see that the energy emitted per second in the beginning in proportional to 3 and at the end is proportional to 2. Therefore the average rate at which energy is being emitted is

$$= 4 \times 5.670 \times 10^{-8}\,\text{Wm}^{-2}\text{k}^{-4} \times (273+27)^3\,\text{k}^3 \times \frac{3+2}{2}\,\text{k}$$

$$\times \left(\frac{5}{100} \times \frac{5}{100}\right) \times 6\text{m}^2$$

$$= 4 \times 5.670 \times 27 \times 2.5 \times 25 \times 6 \times 10^{-6}\,\text{W}$$

$$= 0.230\ \text{J/s}.$$

Total energy radiated by water in cooling through 1°C.

= 125 g × 1 Cal/g –°C × 1°C

= 125 cal

= 125×4.184 J

= 523 J.

Time taken for this cooling $= \dfrac{523\text{J}}{0.230\text{J/s}}$

= 2280s

= 38 min

Problem 3 : The temperature of a body falls from 30°C to 25°C in 5 min. The air temperature is 15°C. Find the temperature after a further 5 minute interval.

Solution : Let the final temperature after a further 5 minute interval be t°C

$$\text{Average temperauture} = \frac{30+25}{2} = 27.5°\text{C}$$

∴ Average excess temperature = 27.5 – 25

= 2.5°C

Average rate of fall of temperature $= \dfrac{5}{5} = 1°\text{C/min}$

Using the relation :

Average rate of fall of temp = k × Average excess temp.

Putting the values in this equation, we get 1°C/min = k × 12.5°C

or $$k = \frac{1}{12.5} \text{ per min.}$$

For next five minutes

$$\text{Average temp.} = \left(\frac{25+x}{2}\right)°C$$

$$\text{Average excess temp.} = \frac{25+x}{2} - 15 = \left(\frac{x-5}{2}\right)°C$$

$$\text{Average rate of fall of temp.} = \left(\frac{25-x}{5}\right)°C/\text{min}$$

$$\therefore \frac{25-x}{5} = \frac{1}{12.5} \times \frac{x-25}{2} = \frac{x-5}{25}$$

$$\text{or} \quad 5(25-x) = x-5$$

$$\text{or} \quad x = 21.7°C$$

Problem 4 : A copper calorimeter weighing 99.0 g is first filled with water and then with kerosene. Time taken to cool from 45°C to 40°C in the two cases is 10 min and 5 min respectively. The mass of water is 100 g and heat of kerosene is 80 g. Calculate the specific heat of kerosene. The specific heat of copper is 0.094 cal/g–°C and room temperature is 35°C.

Solution : All the factors, such as colour, temp., area etc., on which radiation of energy depends are same. The temp. difference in the beginning and at the same is also almost same so the average rate at which energy is radiated by water and calorimeter is equal to

$$= \frac{(99.0 \times 0.094 + 100.0 \times 1) \times 5}{10} \text{cal/min}$$

The average rate at which energy is radiated by kerosene and calorimeter

$$= \frac{(99.0 \times 0.094 + 80.0 \times 5) \times 5}{5} \text{cal/min}$$

Here s is the specific heat of kerosene in cal/g–°C

$$\therefore \quad \frac{(99.0\times 0.094+80.0\times 5)\times 5}{10} = \frac{(99.0\times 0.094+80.0\times 5)\times 5}{5}$$

$$= \frac{9.305+100}{2} = 9.305+80s$$

or $$109.3 = 18.6 + 1605$$

or $$s = \frac{90.7}{160} = 0.57$$

∴ Specific heat of kerosene = 0.57 cal/g–°C

Problem 5 : A furnace is at a temperature of 2000 k. At what wave length would you expect the greatest intensity of radiation? Is it in the visible region ?

Solution : We have the relation

$$\lambda_{m_1} T_1 = \lambda_{m_2} T_2 = b = \text{Constant}$$

where λ_{m_1} and λ_{m_2} are the curve lengths for which the emission is maximum at temperaures T_1 and T_2 respectively.

'b' is a constant known as Wien's constant and its value is 2.898×10^{-3}mk.

Thus we have

$$\lambda_m = \frac{b}{T} = \frac{2.298 \times 10^{-1}\,\text{mk}}{2000\ \text{k}}$$

$$= 1.449 \times 10^{-6}\,\text{m}$$

$$= 1.449 \text{ micron}$$

Visible region lies approximately between 4000 Å (0.4 micron) to 8000 Å (0.8 micron) so this wave length (1.449 micron) is beyond red colour.

Problem 6 : If each square centimeter of the Sun's surface radiates energy at the rate of 1.5×10^3 cal s^{-1} cm^{-2} and Stefan's constant is 5.7×10^{-5} ergs sec^{-1} cm^{-2} deg. abs.$^{-4}$, calculate the temperature of Sun's surface in degrees centigrade, assuming Stefan's law applies to the radiation.

Solution : From Stefan's law, $E = \sigma T^4$

The energy per sq. cm. per sec from the Sun, E

$$E = 1.5 \times 10^3 \text{ calories} = 1.5 \times 10^3 \times 4.2 \times 10^7 \text{ ergs}$$

If T is the absolute temperature of the Sun.

$$1.5 \times 10^3 \times 4.2 \times 10^7 = 5.7 \times 10^{-5} T^4$$

$$\therefore \quad T = \left(\frac{1.5 \times 10^3 \times 4.2 \times 10^7}{5.7 \times 10^{-5}} \right)^{\frac{1}{4}}$$

$$T = \left(\frac{6.3}{5.7} \times 10^{15} \right)^{\frac{1}{4}}$$

$$= 5765k$$

$\therefore$ temperature in °C = 5765–273 = **5492°C**

Problem 7 : Calculate the temperature of the Sun.

Given $s = 1.937$ cal/sq cm/min

$\sigma = 1.37 \times 10^{-12}$ cal/cm^2/min

$a = 4.3 \times 10^5$ miles

$b = 9.28 \times 10^7$ miles.

Solution : Temperature of the Sun,

$$T = \left[\frac{b^2}{a^2} \cdot \frac{s}{60} \cdot \frac{1}{\sigma} \right]$$

$$T = \left\{ \left(\frac{9.28 \times 10^7}{4.3 \times 10^5} \right)^2 \times \frac{1.937}{60} \times \frac{1}{1.37 \times 10^{-12}} \right\}^{\frac{1}{4}}$$

$$= 5780k$$

Problem 8 : The order to magnitude of the energy received from sun at earth's surface is 10^{-1} Joules cm^{-2} sec^{-1}. Calculate the order of magnitude of the total force due to solar radiation on the earth (assumed perfectly absorbing) given earth's diameter = 10^{-7} meter.

[Lucknow 1978]

Solution : The energy recieved on earth's surface per unit area per sec.

$$E=10^{-1}\text{ Joules/cm}^2 = 10^3\text{ Joule m}^{-2}\text{ sec}^{-1}$$

The radiation pressure,

$$P = \frac{E}{C} = \frac{10^3}{3\times10^8}\text{N/m}^2$$

$$= 3.33\times10^{-6}\text{N/m}^2$$

∴ Total force due to solar radiation on earth

$$= P \times \text{surface area}$$

$$= 3.33\times10^{-6}\times4\pi\,r^2$$

$$= 3.33\times10^{-6}4\times3.14\times\left(\frac{10^7}{2}\right)^2 = 1.05\times10^9\text{ N}$$

Problem 9 : Calculate the maximum amount of heat which may be lost per sec by radiation from a sphere of 10 cm in diameter at a temperature of 227°C when placed in an enclosure at a temperature of 27°C. ($\sigma = 5.7\times10^{-12}$ Watt × $\text{cm}^{-2}(°\text{C})^{-4}$

[Kanpur 1978]

Solution : The rate of loss of energy will be maximum if the sphere radiates as black body. By Stefan Boltzmann law, the net energy lost per unit area of the sphere per second is given by

$$E = \sigma(T^4 - T_0^4)$$

Where T and T_0 are the absolute temperature of sphere and the surrounding respectively and σ is Stefan's constant.

Total surface area of sphere = $4\pi\,r^2$

∴ The net rate of loss of heat

from the total surface of the sphere, $Q = E\times4\pi\,r^2$

$$Q = \sigma(T^4 - T_0^4)\times4\pi r^2$$

or $$= \frac{\sigma(T^4 - T_0^4)\times4\pi\,r^2}{J}\text{ calories/sec.}$$

Given : r = 5 am = 5 × 10^{-2} m; T = 227°C = 500k

T_o = 27°C = 300 k; $\sigma = 5.7 \times 10^{-12}$ Watt cm^{-2} $(k)^{-4}$

$= 5.7 \times 10^{-8}$ watt $m^{-2}(k)^{-4}$

J = 4.2 J / cal

$$\therefore\ Q = \frac{5.7 \times 10^{-8}\left[(500)^4 - (300)^4\right] \times 4 \times 3.14 \times \left(5 \times 10^{-12}\right)^2}{4.2}$$

= 23.3cal / sec

Problem 10 : Luminosity of Rigel star in orion constellation is 17000 times that of our Sun. If the surface temperature of the Sun is 6000 k, calculate the temperature of the Star.

[Agra 1978]

Solution :

From Stefan Boltzmann's law the energy radiates per unit surface of the body is given by

$$E = \sigma(T^4 - T_0^4)$$

When T and T_0 are the absolute temperature of body and the surroundings.

In case $T_0 << T$, then we have

$$E = \sigma T^4$$

Now the luminosities of Star and Sun is E_1 and E_2 and their absolute temperatures are T_1 and T_2 respectively then

$$E_1 = \sigma\ T_1^4 \quad \ldots \text{(i)}$$

and $$E_2 = \sigma\ T_2^4 \quad \ldots \text{(ii)}$$

Dividing (i) by (ii), we get

$$\frac{E_1}{E_2} = \frac{T_1^4}{T_2^4}$$

Since $$\frac{E_1}{E_2} = 17000 \quad \text{(given)}$$

and $T_2 = 6000$ k (given)

$\therefore \quad T_1^4 = 17000 \times (6000)^4$ [given Eq (iii)]

or $T_1 = (17000)^{\frac{1}{4}} \times 6000$

$= 68520$k.

Problem 11 : Two ideal black bodies A and B at temperatur 227°C and 327°C respectively are placed in an evacuated enclosure whose walls are blackened and kept at 27°C. Compare the rates of loss of heat. [Agra 1972]

Solution : Energy lost by a body per sec per m^2, E is given by

$$E = \sigma\ (T^4 - T_0^{\ 4})$$ [Stefan-Boltzmann's Law]

Given : $T = 227°C = 500$ k

$T_0 = 27°C = 300$ k [For body A]

$$\therefore \quad E_1 = \sigma\left[(500)^4 - (300)^4\right]$$

and for body B,

$T = 327°C = 600$ k

$T_0 = 27°C = 300$ k

$$\therefore \quad E_2 = \sigma\left[(600)^4 - (300)^4\right]$$

$$= 1215 \times 10^8 \sigma$$

or $$\frac{E_1}{E_2} = \frac{544 \times 10^8 \sigma}{1215 \times 10^8 \sigma} = \frac{544}{1215}$$

Hence the ratio of heat loss rates of A and B is

A : B = 544 : 1215

Problem 12 : A body of mass 10 gm is kept in an enclosure of temperature 27°C. If the temperature of the body is 127°C, its specific heat is 0.1 kilo calorie per kg per deg C and area of emitting surface of the body is 10^{-3} m^2. Find out the rate of cooling of the body. [Kanpur 1976]

Solution : The quantity of energy (Q) radiated by a body with surface area, A, and temperature T k in time, t, is given by

$$Q = A\sigma\left(T^4 - T_0^4\right).t$$

$$\therefore \quad \frac{dQ}{dt} = A\sigma\left(T^4 - T_0^4\right)$$

Since Q = m.s.T.

$$\therefore \quad \frac{d(msT)}{dt} = A\sigma\left(T^4 - T_0^4\right)$$

Hence Rate of cooling of the body $\frac{dT}{dt} = \frac{A\sigma}{ms}\left(T^4 - T_0^4\right)$

$$= \frac{10^{-3} \times \left(\frac{5.72 \times 10^{-8}}{4.2 \times 10^3}\right)\left[(400)^4 - (300)^4\right]}{\left(10 \times 10^3\right)(0.1)}$$

$$= \frac{10^{-3} \times 5.72 \times 10^{-8} \times \left[(400)^4 - (300)^4\right]}{4.2 \times 10^3 \times 10 \times 10^{-3} \times 0.1}$$

$$= 0.23°C\ sec^{-1} \qquad = 0.23\ k\ \ sec^{-1}$$

Problem 13 : A black body with an initial temperature of 300°C is allowed to cool inside an evacuated enclousre surrounded by melting ice at the rate of 0.35°C per sec. If the mass, specific heat and surface area of the body are 32 g, 0.10 cal/gm °C and 8 cm^2 respectively. Calculate the Stefan's constant

[Agra 1977]

Solution : The energy radiated, E, per second per unit area from a body at temperature Tk surrounded by an enclosure at T_0k, is

$$E = \sigma\left(T^4 - T_0^4\right).$$

For a body having surface area, A, the total loss of energy (Q) due to radiation is

$$Q = \sigma.(T^4 - T_0{}^4).A \text{ Joules / sec.}$$

$$= \frac{\sigma.\left(T^4 - T_0^4\right).A}{J} \text{kilo cal / sec} \quad ...(i)$$

The loss of energy due to radiation from a body is also given by

$$Q = \text{m.s. } (\Delta t) \quad [\Delta t = \text{fall in temp. per sec.}]$$

$$= 32 \times 10^{-3} \times 0.1 \times 0.35$$

$$= 1.12 \times 10^{-3} \text{ k cal / sec.} \quad ...(ii)$$

Comparing (i) and (ii)

$$\frac{\sigma\left(T^4 - T_0^4\right)A}{J} = 1.12 \times 10^{-3}$$

or $$\sigma = \frac{1.12 \times 10^{-3} J}{\left(T^4 - T_0^4\right)A}$$

$$= \frac{1.12 \times 10^{-3} \times 4.20 \times 10^3}{\left[(573)^4 - (273)^4\right] \times 8}$$

$$= 5.7 \times 10^{-8} \text{ Joule m}^{-2}\text{sec}^{-1}(k)^{-4}$$

Problem 14 : Calculate the energy radiated per minute from the filament of an incandescent lamp at 2000 k, if the surface area is 5×10^{-5} metres^2 and its relative emittance is 0.85.

[Bhopal 1984]

Solution : We know

$$E = \sigma\left(T^4 - T_0^4\right) \quad \text{[Stefan Boltzmann Law]}$$

if $T >> T_o$, then [T = Temp. of body

T_o = Temp. of sorroundings]

$$E = \sigma T^4$$

For a body having relative emittance, e, we have

$$E = \sigma e \, T^4$$

The total energy radiated by the body per sec

$$= \sigma e \left(T^4 - T_0^4\right) \times \text{surface area}$$

$$E = 5.7 \times 10^{-8} \times 0.85 \times (2000)^4 \times \left(5 \times 10^{-5}\right)$$

Hence the energy radiated by body per minute is equal to

$$= 5.7 \times 10^{-8} \times 0.85 \times (2000)^4 \times \left(5 \times 10^{-5}\right) \times 60$$

$$= 2315 \text{ Joules}$$

Problem 15 : Calculate the surface temperature of the Sun. Given for solar radiations $\lambda_m = 4753$ A° and Wiens constant, b = 2.898 × 10^{-3} meter-kelvin.

[Rewa 1985]

Solution : According to Wiens law

$$\lambda_m . T = \text{constant} = 0.2898 \times 10^{-3} \text{mk}$$

Given : $\lambda_m = 4753\text{A}° = 4753 \times 10^{-10} \text{ m}$

$$\therefore T_0 = \frac{0.2898 \times 10^{-3}}{4753 \times 10^{-10}} = 6097\text{k}$$

Problem 16 : From Wein's displacement law, we have

$$\left[\lambda_m . T = \frac{hc}{5k} = b\right]$$

(i) Calculate the constnat, b, what are the units of b?

(ii) Radiation from moon given $\lambda_m = 4700$ A° and 14 × 10^{-6} m. What conclusions can you draw from this?
[h = 6.6 × 10^{-34} J/s, K = 1.37 × 10^{-23} J/k, E = 3 × 10^8 m/s]

[Lucknow 1984]

Solution : (i) Since

$$b = \frac{hc}{5k}$$

$$b = \frac{6.6 \times 10^{-34} \times 3 \times 10^8}{5 \times 1.37 \times 10^{-23}} = 0.002890 \text{ mk}$$

(ii) $\lambda_m = 4700° A = 4700 \times 10^{-10}$ m,

So we have

$$T_1 = \frac{0.002890}{4770 \times 10^{-10}} = 6166 \text{ k.}$$

when
$$\lambda_m = 14 \times 10^{-6}$$
$$T_1 = \frac{0.002890}{14 \times 10^{-6}} = 270 \text{ k}$$

Conclusions : The first one is the temperature of the Sun (6166k). As the Sun's radiations are reflected from moon's disc so the other one (207 k) is the moon's temperature's.

Problem 17 : The earth recieves 1.90 calories of heat per cm2 minute. Calculate the surface temperture of the sun. Given the distance of earth from the sun = 150 × 10^6 km.

[Stefan's constant 1.37 × 10^{-12} cal cm^{-2} sec^{-1} $degree^{-4}$]

[Meerut 1984]

Solution : For the surface temperature (T) of the sun, we have

$$T = \left[\left(\frac{R}{r}\right)^2 \times \frac{5}{60} \times \frac{1}{\sigma}\right]^{\frac{1}{4}}$$

Given
$$R = 150 \times 10^6 \text{km} = 150 \times 10^9 \text{ m}$$
$$r = \frac{1.39 \times 10^6}{2} \text{km} = 0.695 \times 10^9 \text{ m}$$
$$\sigma = 1.37 \times 10^{-12} \text{ cal m}^{-2} \text{sec}^{-1} \text{degree}^{-4}$$
$$= 1.37 \times 10^{-8} \text{ cal m}^{-2} \text{sec}^{-1} \text{degree}^{-4}$$
$$S = 1.90 \text{calories.}$$

$$T = \left[\left(\frac{150 \times 10^9}{0.695 \times 10^9}\right) \times \frac{1.9 \times 10^9}{60} \times \frac{1}{1.37 \times 10^{-8}}\right]^{\frac{1}{4}}$$
$$= 570\text{k}$$

Problem 18 : The earth receives 2.0 cal cm^{-2} min^{-1} from the Sun. If the angular diameter of the Sun is 32' and it is treated as a black body, deduce its surface temperature. Given $\sigma = 5.7 \times 10^{-12}$ watt $cm^{-2}(k^{-1})$

[Agra 1976]

Solution : T_1 the temperature of the Sun is given by

$$T = \left[\left(\frac{R}{r}\right)^2 \times \frac{S}{60} \times \frac{1}{\sigma}\right]^{\frac{1}{4}}$$

Now $\frac{r}{R} = \frac{\text{Radius of the Sun}}{\text{distance from the earth}}$ = angular radius

Since angular diameter = 32′ (given)

$$\therefore \text{ angular radius, } \frac{r}{R} = 16' = \left(\frac{16}{60}\right)^0$$

$$= \frac{16}{60} \times \frac{\pi}{180} \text{ radians}$$

$$\text{Hence} \quad \frac{R}{r} = \frac{180 \times 60}{16\pi}$$

$$= 2.0 \text{ cal } m^{-2} min^{-1}$$

$$= (2.0 \times 4.2).J \times \left(10^{-2} m\right)^{-2} \times min^{-1}$$

$$= 2 \times 4.2 \times 10^4 \, J\, m^{-2} min^{-1}$$

$$\sigma = 5.7 \times 10^{-12} \text{ watt } cm^{-2}(k)^{-4}$$

$$= 5.7 \times 10^{-8} \text{ watt } m^{-2} min^{-1}$$

Substituting these values, we get

$$T = \left[\left(\frac{180 \times 60}{16\pi}\right)^2 \times \frac{2 \times 4.2 \times 10^4}{60} \times \frac{1}{5.7 \times 10^{-8}}\right]^{\frac{1}{4}}$$

$$= 5803 \text{ k}$$

Problem 19 : Find the pressure exerted by sunlight on earth's surface if the value of solar constant is 2 cal/min cm^2.

[Rohilkhand 1990]

Solution : The radiation pressure, p, is given by

$$P = \frac{E}{C}$$

Given : $E = 2 \text{ cal/min cm}^2$

$$= \frac{2 \times 10^4}{60} \text{cal / sec m}^2$$

$$= \frac{2 \times 10^4}{60} \times 4.2 \text{ Joules / sec m}^2$$

$$= 1.4 \times 10^3 \text{ J / sm}^2$$

Hence the pressure exerted on earth by sunlight, p is

$$p = \frac{E}{C} = \frac{1.4 \times 10^3}{3 \times 10^8}$$

$$= 4.66 \times 10^{-6} \text{N / m}^2$$

Problem 20 : Calculate the number of modes of viration in a 100 cc chamber in the wave length range (i) 5000 to 5002 A° (ii) 8000 to 8005 A°

[Rajasthan 1980s]

Solution : The number of modes per cc in the wave length range λ to λ + dλ

$$= \frac{8\pi}{\lambda^4} d\lambda$$

Hence for a chamber having volume, v, the number of modes in the wave length range λ to λ + dλ is given by

$$N_1 = \frac{8\pi v}{\lambda^4} d\lambda$$

(i) Given v = 1000 cc; $d\lambda = 5002 - 5000 = 2 \text{ A°} = 2 \times 10^{-8}$ cm

$\lambda = 5000 \text{Å} = 5000 \times 10^{-8} \text{cm}.$

$$\therefore N_1 = \frac{8 \times 3.14 \times 100}{(8000 \times 10^{-8})^4} \times 2 \times 10^{-8}$$

$$= 8.03 \times 10^{12}$$

(ii) $\quad v = 100$ cc, $\lambda = 8000$ A° $= 8000 \times 10^{-8}$ cm

$$d\lambda = (8005 - 8000)\text{Å} = 5\text{Å} = 5 \times 10^{-8}\,\text{cm}.$$

$$\therefore\ N_2 = \frac{8 \times 3.14 \times 100}{(8000 \times 10^{-8})^4} \times 5 \times 10^{-8}$$

$$= 3.066 \times 10^{12}$$

Problem 21 : From Planck's law deduce the value of v corresponding to peak of E_v, v curve at 1000 k. In what spectral region does this frequencey lie ?

[Agra, 1972s]

Solution : Plank's radiation law is

$$E_v d_v = \frac{8\,\pi\, h\, v^2}{c^3} \cdot \frac{1}{e^{\frac{hv}{kT}} - 1} dv \qquad \text{...(i)}$$

For maximum value of E_v, differentiating eq (i) w.r.t. v and equating it to zero, we get

$$\frac{dE}{dv} = \frac{8\,\pi\, h}{c^3}\left[3v^2 \cdot \frac{1}{e^{\frac{hv}{kT}} - 1} + v^3 \frac{\left(-\frac{n}{\chi T}\right).e^{\frac{hv}{hT}}}{\left(e^{\frac{hv}{hT}} - 1\right)^2}\right] = 0$$

or
$$= \frac{8\,\pi\, h\, v^2}{e^{\frac{hv}{kT}} - 1}\left[3 - \frac{\left(-\frac{h}{kT}\right).e^{\frac{hv}{kT}}}{\left(e^{\frac{hv}{kT}} - 1\right)^2}\right] = 0$$

or
$$3 - \frac{h\,v}{kT} \cdot \frac{e^{\frac{hv}{kT}}}{\left(e^{\frac{hv}{kT}} - 1\right)} = 0$$

Put $\quad \dfrac{h\,v}{kT} = x$, then the above equation becomes

$$3 - \frac{x.e^x}{e^x - 1} = 0$$

or $$\frac{x - e^x}{e^x - 1} = 3$$

Solving this we get, $x = \frac{h\nu_m}{kT} = 3$

or V_m corresponding to maximum energy of emission i.e. it corresponds to peak of E_v, v curve

Hence $$V_m = \frac{3kT}{h}$$

Given : $T = 1000$ K

$$\therefore\ V_m = \frac{3 \times 1.38 \times 10^{-16} \times 1000}{6.62 \times 10^{-27}} = 6.25 \times 10^{13} S^{-1}$$

If lies on I.R. region of spectrum.

19

Conduction and Convection

Problem 1 : A steady stream of water flowing at the rate of 500 g/min. through a glass tube 30 cm long, 1 cm in external diameter and 8 mm in bore, the outside of which is surrounded by steam at a pressure of 760 mm of mercury, is raised in temperature from 20°C to 30°C as it passes through the tube. Find the conductivity of glass.

Solution : Mass of water flowing through the tube/min = 500 g

Rise of temperature $= 30 - 20 = 10°C$

$= m\,(Q_2 - Q_1)/t$

$\therefore$ Heat gained by water/sec $= \dfrac{500 \times 10}{60}$ cal

Thermal conductivity of glass K $= \dfrac{Q\, ln\frac{r_2}{r_1}}{2\pi \ell\left(Q_3 - \frac{Q_1+Q_2}{2}\right)}$

Where Q_3 in the temperature of steam at 760 mm pressure and 100°C.

$$\therefore K = \frac{\frac{250}{3}\, ln\, \frac{10}{8}}{2\pi \times 30\left(100 - \frac{30+20}{2}\right)}$$

$$= \frac{250}{3} \times \frac{2.303 \log 1.25}{2\pi \times 30 \times 75}$$

$$= 0.001315$$

Problem 2 : A vessle having an external surface area of 250 sq cm and walls 2 mm thick, is filled with lumps of ice mixed with water and then palced in a water both maintanied at 95°C. How many graims of ice will melt per minute when conditions become steady ?

[Latent heat of fusion of ice = 80 cal/gm; thermal conductivity of material of vessel = 0.002 C.G.S units]

Solution : In the steady state the quantity of heat passing through the walls of the vessle is given by Q/sec = K.A × temperature gradient. Here we have; K = 0.002 C.G.S. units

$$A = 250 \text{ sq. cm}$$

$$\text{Temp. gradient} = \left(\frac{95-0}{0.2}\right)°C/cm$$

$$\therefore Q/sec = \frac{0.002 \times 250 \times 95}{0.2} cal/sec$$

$$\therefore Q/mm = \frac{0.002 \times 250 \times 95 \times 60}{0.2} cal/min$$

$$= 14250 \ \ cal/min$$

But latent heat of fusion of ice = 80 cal/gm

$$\therefore \text{mass of ice metted/min} = \frac{14250}{80}$$

$$= 178.125\ g$$

Problem 3 : One end of a cylindrical metal rod, 2 mm in diameter, is screwed into copper block weighing 132 g. The other end projects into a steam jacket and can be maintained at a temperature of 100°C. The length of the exposed rod between the steam jacket and the face of the block is 10 cm. When the temperature of the apparatus is 10°C, steam is turned into the jacket and the initial rate of rise of temperature of the copper black is observed to be 1°C/min. Assuming that no heat escapes from the surface of the rod, find

(a) The number of calories of heat passing through the rod from the steam jacket in 1 sec.

(b) The coefficient of thermal conductivity of the material of which the rod is made.

[Sp. heat of copper = 0.09 cal/g/°C].

Solution : (a) The heat conducted through the rod initially raises the temperature of the copper block at the rate of 1°C per min or $\frac{1}{60}$ °C per sec.

$\therefore$ Heat supplied to copper block/sec = Q/sec

$$= 132 \times 0.09 \times \frac{1}{60} \text{cal / sec}$$

$$= 0.198 \text{cal / sec}$$

Thus heat passing through the rod from steam jacket in one second = 0.198 cal.

(b) The temperature gradient across the exposed rod when the steam is turned on $= \frac{100-10}{10} = 9°\text{C / cm}$

The area of cross = section of the rod = $\pi r^2 = \pi \times 0.1^2 \text{cm}^2$

Since we have already calculated in part (a) that the heat passing through the rod/sec = 0.198 cal

$$\therefore \quad 0.198 = k \times \pi \times 0.1 \times 9$$

or
$$k = \frac{0.198}{3.14 \times 0.01 \times 9} = 0.70 \text{cal / cm}^2 \text{ / s}$$

Problem 4 : The temperature inside a bolier is 105°C. The wall of the boiler is 2 cm thick and is lagged with 4 cm thickness of a material whose termal conductivity is $1/9$ of that the material of the boiler. When in the steady state the temperature of the outside surface of the lagging in contact with the air is 10°C. What is the temperature of the common surface of the boiler and the lagging?

Solution : Let the temp. of common surface of boiler and lagging = t°C

Then temperature gradient across the wall of the boiler

$$= \left(\frac{105-t}{2}\right) \text{C / cm}$$

and temperature gradient across the lagging $= \left(\frac{t-10}{4}\right) °\text{C / cm}$

If A is the area of the wall as also the area of the lagging material and k and $\frac{k}{9}$ are respective thermal condutivities, then the quantities of heat passing per second through the wall and the lagging in the steady state are respectively.

$$k - A. = \left(\frac{105-t}{2}\right) \text{ and } \frac{k}{9}.A.\left(\frac{t-10}{4}\right) \text{cal / sec}$$

But in steady state both these must be equal. Thus

$$k.A.\left(\frac{105-t}{2}\right) = \frac{k}{9}.A\left(\frac{t-10}{4}\right)$$

$$\therefore \frac{105-t}{2} = \frac{t-10}{9\times4}$$

$$\therefore 19t = 1900$$

$$\text{or } t = 100°C$$

Problem 5 : Two equal plates with thermal conductivities k_1 and k_2 are joined together to form a single plate of double thickness. Calculate the equivalent thermal conductivity of the composite plate so formed.

[Gorakhpur 1983]

Solution : Let A be the surface area and x the thickness of each plate. Also let θ_1 and θ_2 be the temperatures of the end faces and θ be the temperature of the surfaces in contact. Under steady conditions, the rate of heat flow through both plates will be the same, hence.

$$\theta = \frac{k_1.A(\theta_1-\theta)}{x} = \frac{k_2.A(\theta-\theta_2)}{x} \quad \text{...(i)}$$

$$\text{or } k_1\theta_1 - k_1\theta = k_2\theta - k_2\theta_2$$

$$\text{or } k_1\theta_1 + k_2\theta_2 = \theta(k_1+k_2)$$

$$\theta = \frac{k_1\theta_1 + k_2\theta_2}{k_1+k_2} \quad \text{...(ii)}$$

If the thickness of composite plate is 2x and k is its equivalent thermal conductivity, then rate of heat flow through it

$$\theta = \frac{kA(\theta_1-\theta_2)}{2x} \quad \text{...(iii)}$$

From equations (i) and (iii), we get

$$\frac{k_1A(\theta_1-\theta)}{x}=\frac{kA(\theta_1-\theta_2)}{2x}$$

or $$k=\frac{2k_1(\theta_1-\theta)}{\theta_1-\theta_2}$$

or $$k=\frac{2k_1(\theta_1-\theta)}{\theta_1-\theta_2}$$

or $$k=\frac{2k_1}{(\theta_1-\theta_2)}\left[\theta_1-\frac{k_1\theta_1+k_2\theta_2}{k_1+k_2}\right] \qquad [\because \text{Eq (ii)}]$$

$$=\frac{2k_1k_2}{k_1+k_2}$$

Problem 6 : An insulating wall of area 4 square meters is made from cork 16.5 cm thick protected externally by 12.5 cm brick and lined on the inside with 10 cm wood. If the exposed surface of the wood is maintained at –5°C and the outer surface of the brick at 20°C, calculate

(i) The rate of energy transfer through the wall,

(ii) The interface temperatures.

Given : Thermal conductivity

For brick = 2.5×10^{-3}, For c or k = 1.1×10^{-4}

and For wood = 4×10^{-4} C.G.S. units. [Agra 1970]

Solution :

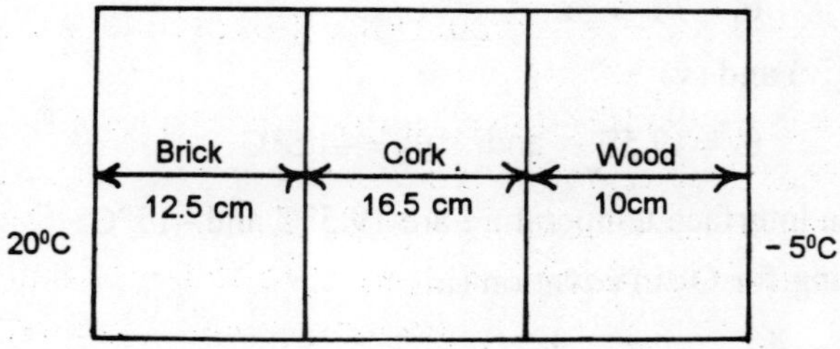

fig.

The rate of flow of heat through any surface is given by

$$Q=\frac{kA\left(\theta-\theta^1\right)}{x}$$

Here $A = 4 \times 10^4$ Sq. cm and for brick

$k = 2.5 \times 10^{-3}$ cal sec^{-1} cm^{-1} c^{-1}

Hence through bricks, rate of heat flow

$$Q_1 = \frac{2.5\times10^{-3}\times4\times10^4(20-\theta_1)}{12.5} = 8(20-\theta_1) \quad ...(i)$$

Similarly through cork and wood, the rate of heat flow are respectively

$$Q_2 = \frac{1.1\times10^{-4}\times4\times10^4(\theta_1-\theta_2)}{16.5} = \frac{4}{15}(\theta_1-\theta_2) \quad ... (ii)$$

$$Q_3 = \frac{4\times10^{-4}\times4\times10^4(\theta_2-(-5))}{10} = \frac{8}{5}(\theta_2+5) \quad ... (iii)$$

In the steady state the rate of flow through all the surfaces is the same. Hence from (i) and (ii)

$$8(20-\theta_1) = \frac{4}{15}(\theta_1-\theta_2)$$

$$31\theta - \theta_2 = 600 \quad ...(iv)$$

Also from (ii) and (iii)

$$\frac{4}{15}(\theta_1-\theta_2) = \frac{8}{5}(\theta_1+5)$$

$$\theta_1 - 7\theta_2 = 30 \quad (v)$$

Solving (iv) and (v)

$$\theta_1 = 19.5° \quad \text{and} \quad \theta_2 = -1.5°C$$

Hence the interface temperature are 19.5°C and –15°C

Substituting for Q_2 in equation (iii)

$$Q_3 = \frac{8}{5}(-1.5+5) = \frac{8}{5}\times3.5$$

$$= 5.5 \text{cal / sec}$$

Thus the rate of energy transfer through the wood (or wall as a whole) is 5.5 cal/sec.

Problem 7 : A bar of length 40 cm and cumform cross-section 5 cm^2 consists of two halves. AB of copper and BC of iron welded together at B. The end A is maintained at 200°C and end C at 0°C and the sides are thermally insulated. Find the rate of flow of heat along the bar when the steady state has been reached.

(Thermal conductivities are : copper 0.9; iron 0.12)

Solution : Under steady conditions, the rate of heat flow through both sections of the bar will be same i.e.

$$\frac{\theta}{t} = \frac{k_1(200-\theta)}{d} = \frac{k_2(\theta-0)}{d}$$

$$0.9 \times (200-\theta) = (\theta-0)$$

$$\theta = 176.4°c$$

Putting the value, $\theta = 176.4°C$, $A = 5\ cm^2$ and $d = \frac{40}{2} = 20$ in Eq (i) we get rate of flow of heat along the bar

$$\frac{Q}{t} = \frac{0.9 \times 5 \times (200-176.4)}{20}$$

$$= 5.31 cal/sec$$

Problem 8 : The thickness of ice on a lake is 5.0 cm and the temperature of air is –20°C. Calculate hav long will it take for the thickness of ice to be doubled. Thermal conductivity of ice = 0.005 cal cm^{-1} sec^{-1} $°C^{-1}$; density of ice = 0.92 g/cc and latent heat of ice = 80 cal/gm.

[Agra 1971]

Solution : We know

$$dt = \frac{x\ell L}{k\theta} dx$$

So the time needed for thickness to increase from x_1 to x_2 is

$$t = \frac{\ell L}{k\theta} \int_{x_1}^{x_2} x\, dx$$

$$\text{or} \qquad t = \frac{\ell L}{k\theta}\left(\frac{x^2}{2}\right)_{x_1}^{x_2} = \frac{\ell L}{2k\theta}\left(x_2^2 - x_1^2\right)$$

Substituting the given values, we get

$$t = \frac{0.92 \times 80}{2 \times 0.005 \times 20}\left(10^2 - 5^2\right)$$

$$= 27600\text{sec} = 7 \text{ hour } 40 \text{ minutes}$$

Problem 9 : In an Ingen-Hausz's experiment, three identical rods of different materials are coated with wax and placed with one end inserted in an oil-bath containing hot water, when steady state is reached, wax on the rods is formed melted upto length l_1, l_2 and l_3. Compare the conductivity of the materials of the rods.

[Calcutta 1986]

Solution : If the rod is so long that its free ends may be considered to be at the temperature of the surroundings. If θ is the excess temperature of the heated ends, then for temperature distribution along each bar is given by

If the m.p. of wax of θm, then $\theta = \theta_m$ at $x = l_1$ for bar *1* at $x = l_2$ for bar 2 and $x = l_3$ for bar 3.

$$\therefore \qquad \text{for bar} 1, \theta_m = \theta_0 . e^{-\mu_1 . \ell_1}$$

$$\text{for bar } 2, \theta_m = \theta_0 . e^{-\mu_2 . \ell_2}$$

$$\text{for bar } 3, \theta_m = \theta_0 . e^{-\mu_3 . \ell_3}$$

$$\therefore \quad \mu_1 \ell_1 = \mu_2 \ell_2 = \mu_3 \ell_3$$

$$\text{or} \qquad \mu_1^2 \ell_1^2 = \mu_2^2 \ell_2^2 = \mu_3^2 \ell_3^2$$

$$\text{But} \quad \mu_1^2 = \frac{E\,\ell}{k_1 A},\ \mu_2^2 = \frac{E\,\ell}{k_2 A},\ \mu_3^2 = \frac{E\,\ell}{k_3 A}$$

Since A for all the rods is same

$$\frac{E\,\ell}{k_1 A},\ \ell_1^2 = \frac{E\,\ell}{k_2 A},\ \ell_2^2 = \frac{E\,\ell}{k_3 A}\ell_3^2$$

$$\text{or} \qquad \frac{\ell_1^2}{k_1} = \frac{\ell_2^2}{k_2} = \frac{\ell_3^2}{k_3}$$

$$\text{or} \qquad k_1 : k_2 : k_3 = \ell_1^2 : \ell_2^2 : \ell_3^2$$

i.e. the conductivities of the different rods vary as the square of the lengths upto which wax has melted on them.

Problme 10 : A rubber tube of length 20 cm through which steam at 100°C is passing, is immersed in a calorimeter whose water equivalent is 15g and which contains 300 gm of water at 16°C. The temperature of water rises at the rate of 2°C per minute. If the outer and inner diameters of the tube are 1.0 cm and 0.6 cm respectively, calculate the thermal conductivity of rubber.

[Gorakhpur 1976]

Solution : For a tube of length, l_1 and internal radii, r_1 and r_2 the radial rate of heat flow is given by the equation.

$$\frac{dQ}{dt} = \frac{2\pi k\ell(\theta_1 - \theta_2)}{\left(\frac{r_2}{r_1}\right)} = \frac{2\pi k\ell(\theta_1 - \theta_2)}{2.303 \log\left(\frac{r_2}{r_1}\right)} \qquad \text{...(i)}$$

where θ_1 and θ_2 are the inside and outside temperatures of the tube. If the temperature of m gm of water contained in the calorimeter having water equivalent W increases by $\Delta\theta$, then

$$\frac{d\theta}{dt} = (m + W)\Delta\theta \qquad \text{....(ii)}$$

Equating (i) and (ii), we get

$$\frac{2\pi k\ell(Q_1 - Q_2)}{2.303 \log\left(\frac{r_2}{r_1}\right)} = (m + W)\Delta\theta \qquad \text{....(iii)}$$

$$\text{or } k = \frac{(m + W)\Delta\theta \times 2.303 \log\left(\frac{r_2}{r_1}\right)}{2\pi\ell(\theta_1 - \theta_2)} \qquad \text{....(iv)}$$

Given : $(m + W) = (300 + 15)g = 315g$

$$\Delta\theta = 2°C/\text{minutes} = \frac{1}{30} °C/\text{sec}$$

$$r_2 = 1\text{cm};\ r_2 = 0.6\text{cm}$$

$$\theta_1 = 100°c;\ \theta_2 = 16°\theta$$

Substituting the given values in eq. (iv), we get

$$k = \frac{315 \times \left(\frac{1}{30}\right) \times 2.303 \times \log\left(\frac{1}{6}\right)}{2 \times 3.14 \times 20 \times (100-16)}$$

$$= \frac{10.5 \times 2.303 \times 0.2217}{2 \times 3.14 \times 20 \times 84}$$

$$= 5.1 \times 10^{-4} \text{ cal cm}^{-1}\text{sec}^{-1}{}^{\circ}\text{C}^{-1}$$

Problem 11 : Two thin concentric spherical shells of radii 5 cm and 10 cm respectively have their annular cavity filled with charcoal powder. When energy is supplied at the rate of 10.5 watts to a heater at the centre, a temperature difference of 60°C is set up between the shells. Find the thermal conductivity of charcoal.

(J = 4.2 J/cal)

[Kanpur 1979]

Solution : We know

$$k = \frac{\theta(r_2 - r_1)}{4\pi\, r_1\, r_2 (\theta_1 - \theta_2)}$$

Where K = Thermal conductivity of charcoal.

Given : $Q = 10.5$ Joule / sec $\quad [\because\ 10.5 \text{ watt} = 10.5\ \text{J/s}]$

$$= \frac{10.5}{4.2} \text{cal/sec} \qquad [\because\ 1 \text{ cal} = 4.2\text{J}]$$

$$r_2 = 10 \text{ cm};\quad r_1 = 5 \text{ cm}$$

$$(\theta_1 - \theta_2) = 60°C$$

Substituting these values in the above equation,

$$k = \frac{\frac{10.5}{4.2}(10-5)}{4 \times 3.14 \times 10 \times 5 \times 60}$$

$$= 3.3 \times 10^{-4} \text{cal cm}^{-1}\log^{-1}\sec^{-1}$$

Problem 12 : In a periodic flow of heat along an iron bar, the periodic time is 4 minute. If the tempeature maximum travels 6 cm in one minute, calculate the thermal conductivity of iron. Density of iron = 7.8 g/cm^3; specific heat of iron = 0.11 cal/gm °C.

[Rohilkhand 1986]

Solution : The velocity of heat wave, v, along the iron bar is given by

$$v = \sqrt{2wh} = \sqrt{2 \times \frac{2\pi}{T} \times \frac{k}{\ell.s}}$$

or $$v^2 = \frac{4\pi k}{T\ell s}$$

Given : $v = 6 \text{ cal / minute} = 0.1 \text{ cm / sec}$

$T = 4 \times 6 = 240 \sec$

$S = 0.11 \text{cal / gm}° \text{C}, \ell = 7.8 \text{g / cm}^3$

Therefore thermal conductivity of iron,

$$k = \frac{v^2.T\ell s}{4\pi}$$

$$= \frac{(0.1)^2 \times 240 \times 7.8 \times 0.11}{4 \times 3.14}$$

$$= 0.1639 \text{cal cm}^{-1}\sec^{-1}{}^\circ\text{C}^{-1}$$

❑❑❑